Praise for the Olive series

'Drinkwater has a sharp eye for character, and the people who populate *The Olive Route* will not disappoint her fans'

Independent on Sunday

'Drinkwater is a rare writer who tackles other people brilliantly . . . Vibrant, intoxicating and heart-warming'

Sunday Express

'A storyteller of great economy and deftness and one who, in telling tales, captures the Midi effectively. There are few better portraits of the different worlds – farmers and water-diviners, crooks and Euro-trash, Algerians and wearying officialdom – that intersect to create the Côte d'Azur'

Daily Telegraph

'The new leader of the pack' *The Times*

'Charming and well written. As the olives ripen to a luscious, dark-purply black, the reader is drawn into the peculiarities and joys of Provençal life' *Daily Mail*

'A spellbinding memoir and a must for anyone who dreams of moving to a kinder climate and starting a new life'

Choice

'A lot of its attraction for me stems from the honesty of the writing, the attention to detail, the intensity of the writer's relationship with the land . . . Above all you feel like you're observing a very intimate moment in someone else's life . . .

Waterstones

Best known for her role as Helen Herriot in BBC Television's *All Creatures Great and Small* (for which she was awarded The Variety Club Television Personality of the Year Award), Carol Drinkwater has enjoyed a long and distinguished career as both an actress and writer.

During her acting career, she has worked in film, television and theatre. Her credits include working with Laurence Olivier at the National Theatre, Stanley Kubrick on *A Clockwork Orange* and Max Von Sydow, whom she played opposite in the film *Father*, for which she won the Critics Circle Award for Best Film Actress.

Carol Drinkwater has written novels for both adults and children, including *An Abundance of Rain, Akin to Love* and *Mapping the Heart*. Her bestselling children's novel, *The Haunted School*, has sold more than 170,000 copies and was made into a film and television series which won the Chicago Film Festival Award for children's films. She is also the author of the best selling memoirs *The Olive Farm, The Olive Season, The Olive Harvest, The Olive Route* and *The Olive Tree*. Visit her website at www.caroldrinkwater.com

By Carol Drinkwater

TRAVEL/BIOGRAPHY

The Olive Farm
The Olive Season
The Olive Harvest
The Olive Route
The Olive Tree
The Illustrated Olive Farm
Return to the Olive Farm

FICTION

An Abundance of Rain
Akin to Love
Mapping the Heart
Because You're Mine

YOUNG ADULT FICTION

The Haunted School
Molly
Molly on the Run
The Hunger: Ireland 1845–1847
Twentieth-Century Girl, London 1899–1900
Suffragette, London 1909–1913
Wow! 366 (contributor)

YOUNG ADULT NON-FICTION

Crossing the Line: Young Women and the Law

Return To The Olive Farm

CAROL DRINKWATER

PHOENIX

A PHOENIX PAPERBACK

First published in Great Britain in 2010
by Weidenfeld & Nicolson
This paperback edition published in 2011
by Phoenix,
an imprint of Orion Books Ltd,
Orion House, 5 Upper Saint Martin's Lane,
London, WC2H 9EA

An Hachette UK company

3 5 7 9 10 8 6 4 2

A CIP catalogue record for this book
is available from the British Library.

ISBN978-0-7538-2681-2

Typeset by Input Data Services Ltd, Bridgwater, Somerset

Printed and bound by CPI Group (UK) Ltd, Croydon CR0 4YY

www.orionbooks.co.uk

To Alan Samson, my editor.
Thank you for a decade of Olive opportunities.
Your passion for language and literature
has been an inspiration.

Acknowledgements

I found this book particularly challenging to write. Of the four books set on the farm, it is the only story I was actually living through while collating the material and I worried that events might not turn out as I intended and that my next few chapters would be ruined or would not happen. The character of 'Luke' is a prime example of such a setback. So, to those who were around me during this process I give very special thanks. Michel, of course, the most loyal and supportive of partners, as well as my mother, my uncle and many other visiting family members, I thank you all profoundly for your patience and your faith.

Professionally, my thanks as always go to the editorial team at W & N: Alan Samson and Lucinda McNeile, as well as fine copyediting from Richard Collins.

Lisa Ginnane who looks after my website, Helen Richardson and her army of publicists, most recently and excellently Sophie Mitchell and Elizabeth Allen. Let me not forget the marketing and sales departments for both hardback and paperback. All round, a very supportive ensemble. Thank you.

Also, to my agent, Jonathan Lloyd and his team at Curtis Brown.

Most importantly, to the sales department, booksellers, and to you, the book buyers, my readers – without you, none of this would be possible. Huge thanks.

'We ask the angels to hold us steady while we behead our fathers, flatten our people, sell our daughters and hang ourselves. And yet every now and then you pluck a daisy, and it breaks your heart.'

Albert Jacquard

'No man is an Island entire of itself; every man is a piece of the Continent, a part of the main: if a clod be washed away by the sea, Europe is the less, as well as if a promontory were, as well as if a manor of thy friend's or of thine own were; any man's death diminishes me, because I am involved in Mankind; And therefore never send to know for whom the bell tolls; It tolls for thee.'

John Donne

1

I heard the thunderous crash while I was ironing, folding tablecloths, bedlinen, napkins by the score.

'Five bottles of white wine smashed!' a female voice shouted, followed by 'Help!' and then my name. It was Bridget, an Anglo-Irish friend from Cannes who had offered to come up and lend a hand with preparations. I hurried from the second of the two stables where we kept washing machine, ironing board and wicker baskets spilling over with dirty clothes, rather grandly referred to as the 'laundry room', to find out what had happened.

'I'm soaked!' she cried.

The bottles had been stored on top of the fridge in the garage because there was no space left within, nor in any of the other fridges belonging to this farm. This one, a rather decaying specimen, was really only fit for beverages and it was stuffed to bursting with bottles of rosé, soft drinks, water and whites. I went in to take a look. Seeping liquid and shards of green glass greeted me. Several chunks of plaster had fallen away from the ceiling, landed on the wine and sent the bottles skittling to the ground. My friend's T-shirt was splashed and stained.

'Oh, sorry! Give it to me. I'm just about to do another load. Borrow one of mine, why not? And I'll clean this mess up.'

Bridget smiled, shook her head. 'Don't worry about it. *Courage*,' she said. 'I best be off to cook himself some lunch and change my clothes! I'll see you later.'

'Thanks for helping.' I kissed her on both cheeks and waved as she wheeled off down the drive.

Where were the others, I wondered. Quashia, our Algerian-Berber gardener, had taken the weekend off and gone to Marseille with a carload of his woolly hatted companions. Michel, with the first of our house guests, had driven to the supermarket hours earlier to do a 'big shop'. I glanced at my watch. It was close to two. The dogs were sleeping beneath the trees, out of the heat. Summer had suddenly swooped in, threatening thunder but never delivering. Besides myself, only the cicadas were busy. I traipsed off to find the dustpan and brush, doubting that we would ever be ready for this upcoming event. Commencing the following Tuesday – today was Saturday – arriving over the course of five days, we were expecting one hundred and twenty-three guests. Michel had decided that a welcome home and belated birthday party for Carol was in order. I cleared up the glass, poured the shards into a box in the boot of my car, to dispose of at the bottle bank down near the gypsy settlement, painstakingly sponged up the three and three-quarter litres of wine, gave the cement floor a brisk mop, took one look at the shambles this garage had become during my absence and retired to the cool white walls within the farmhouse, within my den, where I settled back into an African planter's chair, one of a pair I had found in a *brocante* in the fourth *arrondissement* in Paris some years back.

I had been home almost three weeks and was still in limbo, still unpacking, still attempting to heave myself into the swing of things here at Appassionata, our olive farm.

Silently, I was missing so much. I closed my eyes and listened to the cicadas, to the fan overhead whirring in whispers, transporting me back to foreign lands, to stark, evocative rock formations, sea-fringed, shadowy silhouettes looming into the midday light.

Gone were those scrub-faced mountains of Lebanon with olive trees growing at every turn of the eye, gracefully blotting out the intolerable heat; the ancients and juniors at every step along my way, in every Mediterranean land into which I had ventured. One foot off the bus diving directly into groves, cracked earth beneath my boots, picnicking with nature and birdsong; lizards; solitary walks in the mountain ranges encircling the moist lips of the Mediterranean; surprise encounters with strangers. I was home now and removed from remote backwaters, the villages nestling within hilly enclaves where sonorous church bells tolled the long, slow hours; I had left behind me remote Islamic lands; no more perambulating along streets where the glinting or frowning eyes of the women dressed in black, be they Catholic or Muslim, stared upon me. I felt such a distance now from the war zones, the deprivations, the razed groves of Palestine, the pinched faces of farmers deprived of a living, the fragrant citruses in Sicily encircling classical ruins. And what of the classical ruins rising up out of the deserts of Syria, Libya, of those unexpected gems in Algeria? And what of the hard-nosed oil industry in Spain? Gone was the time warp – the hours of every day spent in past millennia – I had been returned to the twenty-first century – and gone was the freedom to go wherever I pleased, to change my mind at a second's notice without a word or thought to anyone. Such a luxury it had been, a previously untapped freedom, an untapped loneliness as well from time to time.

Less than three weeks earlier, I had deposited my weather-beaten backpack on to the bed – my bed, *our* bed, our capacious space, full of love and secrets, with its inviting mattress and lavender-sweet sheets. This was not some matchbox offering in an unfamiliar room where I was intending to pass a night, or to lay my head to rest for an indefinite, unquantifiable period of time. I was home. I had slipped off my well-worn trekking boots, scuffed but still sturdy, and stowed them at the back of the wardrobe. I would not be needing them again for a while.

Home.

And Michel had decided that it was party time. Perhaps I was not quite ready for an onslaught of festivities? Still, the arrival of friends was always a delight and many I had not seen in a long while. Most were flying in, from countries near or far. It was the perfect moment to introduce Michel to a handful of the many new friends I had made during my travels and Michel wanted us to celebrate the fact that I had actually completed the journeys and found my way back, safely.

After sixteen months alone on the road – aside from one or two short hops back here – I had closed the door on my travels and returned. Journey's end. The quest I had embarked upon – oh, so long ago, it seemed to me now that I had finally set foot on our own dry terraces – had been accomplished, if not completed. Quite possibly my searches would never be completed, that much I knew now. And this quest had sunk beneath my skin; it was a part of me. Much of what I had set out to discover remained undiscovered, and a great many questions and thoughts I had never even considered had been posed, thrown at me, made manifest. I had returned with a million faces in my head, a thousand smiles, beckonings and wavings and they continued to call to me. I had been welcomed into the homes of many a stranger, made bonds, won friends, no enduring enemies as far as I was aware, and had returned to this old farm a different person. I had not conquered, for conquering had not been my objective, but I had penetrated and much had been set before my eyes, revealed.

I felt new, newly born, yet as old as the limestone rocks themselves.

And because I am who I am, the material – piled high beside me now on my desk, alongside my camera – these myriad experiences begged to be assimilated, leavened, shaped into material. Not for nothing . . . There were thousands and thousands of photographs to sift through, dozens of notebooks to reread, conversations and ideas in my head that had been earmarked for notation, consideration . . . And the added delight: these kept me there, on my travels, prevented me from settling. Recalled me at a moment's notice, and when I least expected it, I was gone. Like a puff of smoke, I had disappeared off again out through the wardrobe of my imagination, my memory, travelling through time and space, flying through the souk-like alleys of my mind.

Our old blue bus was wheezing up the drive. Michel gave the toot-toot that signalled his return and I lifted myself out of the chair and went down to greet him and the two guests already installed in the house. He stepped out and gave me a hug. I knew that he was at peace now that I was in France and not some place where he needed to feel concerned. Perhaps it had strained him more than he had expressed.

'Hans and Sabine will unload. How are the preparations coming along?'

'Garage ceiling still collapsing, smashed several bottles of party wine.'

'Oh.' My husband made an immediate detour. I followed and we stood at the door's entrance, side by side, brushing comfortably against one another, looking in.

'I fear we can't leave this much longer,' he muttered. 'I'll get some quotes in after the party. Now, lunch!'

How will we fund this, I wondered. Most of my resources had been sunk into my journeys and Michel's documentary films were not exactly super-lucrative.

'Have you managed to contact Marie-Gabrielle and François?'

François and his warm-hearted wife, Marie-Gabrielle, were apiarists and they had been instrumental in the correlation of my travels through Algeria. She in particular. They had introduced me to a network of very resourceful beekeepers who, during weeks of suicide bombings and deaths in that confused Maghrebian territory, had protected me and offered refuge. It seemed fitting that they should be on the guest list of our summer celebrations, particularly given that none of the Algerians themselves had been able to obtain exit visas.

Michel, strides ahead of me, loaded down with a carton of foodstuffs, shook his head and called back over his shoulder. '*Non.*'

'I'll try them again now.'

'After lunch, please. Let's eat!'

When lovers are reunited after a time apart, there can be a certain shyness in their coming together again; a period of readjustment is frequently required. A first overwhelming rush of passion, of satisfaction – ah, such forgotten completeness, integration, the slaking of the thirst of desire, and then what? One step back, and at that moment comes the realisation that the love you have held so dear is not entirely the reality. Or, rather, the reality does not quite match up to the image you have cherished and carried so tenderly in your heart and head.

And so it was with Appassionata and me. Not so with my husband, Michel, because we had whenever possible stayed in daily contact, nattering on the telephone, sending hugs via emails, brief visits to one another, stolen kisses not specified on the itinerary. Our love had evolved along with the journey as far as I was aware, but our farm had travelled with me only in my heart; in my imaginings and memory, a cracked snapshot as a reminder stuffed into a back pocket. Equally, the farm, its lands and vegetation, had not stood still awaiting my return. France had not stood still. If I had changed, so too had my home and the world in which it flourished. My place, my position on the farm, required renegotiating, re-establishing in the light of new awarenesses, new discoveries made throughout my travels.

The first shock of my homecoming had been the news about the loss of Bassett, our beloved little hunting dog, who, during my peregrinations, had been lethally poisoned by slug pellets, a poison almost certainly intended for the invasive and destructive families of wild boar which haunt our tiny arborous commune. The pellets had been laid out, cunningly wrapped and hidden within food, as a trap on a neighbour's estate, but, tragically, they were discovered and consumed by our efficient little poacher.

The second shock, which came that afternoon when I lifted the phone and punched out the digits that fletched me to the mountains, was the fate of 'our' bees, the fourteen hives of *Apis mellifera*, the European honeybee, that had been wintering in our grounds for a number of years.

The hives had not belonged to us but to the couple I was now trying to contact. They were living in the Alpes-Maritimes heights, at chestnut-tree level, where during wintertime they were frequently snowed in and the temperatures plummeted to levels that were too cold for their honeybees.

The phone began to ring. Michel and I had left several messages over the past two weeks inviting them to the shindig, but none had been replied to.

Honeybees do not die off or hibernate in winter. They reduce their activities and they cluster, clinging tightly together on the combs. The outer bees form a protective insulation ensuring consistent temperatures within. The supreme concern of any honeybee colony is the health and comfort of its queen and, most importantly, the rearing of the brood. As the supplies of pollen and honey diminish and outside temperatures drop, the bees huddle together more tightly, embracing their royal lady, keeping her and the young warm, maintaining the hive at thirty-five degrees Celsius. The bees rely on honey and pollen stores, collected throughout the summer months, for their survival. The colder the outdoor temperatures, the more honey they require.

'Allo?'

'Marie-Gabrielle? *Bonjour, c'est Carol.*'

It was sound logic for a keeper to select a winter placement for his hives that was not too cold. The less honey consumed by the bees, the more remained for the hive-owner and customers. It made sense, then, for our apiarist living at an altitude of 1600 metres above sea level, to choose Appassionata for his 'little girls'. Installing honeybees on our farm had long been an ambition of ours, but, due to our peripatetic lifestyles, taking on such a responsibility ourselves would have been reckless. Sharing the hives of François and his wife, Marie-Gabrielle, had struck us all as the ideal solution.

'Mais, Carol, bonjour, ma chérie, comment tu vas?'

'Very well, thank you, settling back into French life. I was wondering if you had received our messages?'

It had toppled over into June by the time I had ascended Italy, passing through its northern cities, overnighting in Florence, until I reached Milan where I met up with Michel and whence we returned home together. Then, my first sighting of Appassionata in many months. The beehives, as we ascended our winding asphalt drive, were absent from their habitual niche in among the citrus trees, but that was as I had expected. Our apiarists descended from their hand-built chalet every year in early spring, March or the beginning of April, and loaded the hives on to a trailer to transport the 'little girls' to summer pastures, an apian transhumance. Transhumance, the moving of livestock, is a very ancient agricultural tradition. Usually, their first port of call after a winter with us would be a shaded retreat within the lower regions of the southern Alps, a couple of hours' ascent into the hinterland. Frequently, this couple chose to settle 'our' hives in the Mercantour National Park where the honeybees could gorge themselves on the intensely coloured wild rhododendron flowers. Wild rhododendron honey was their speciality, and from the plant's point of view it thrived better when it was pollinated externally by bumble or honeybees.

I had silently noted the empty lemon terraces, but thought nothing of it. 'They are off summer pasturing' was my conclusion. I was still smarting at the news of our hunting hound, Bassett. His death had been a tragic accident, but I was angry just the same. It was one more life lost due to the misuse – overuse – of pesticides and I wanted to lay it squarely at the feet of the chemical multi-nationals. Michel had communicated no updates to me about the bees. The news about our dog was, he judged, sufficient to reduce my homecoming to a bittersweetness.

'Yes, we received the messages, thank you.'

Beekeepers, in my experience, are not always the easiest of folk to contact. Ours had an answering machine and an email address, but they seemed to check them very irregularly.

'I hope you will be joining us,' I said, after outlining the occasion once more.

'How we would love to see you, to come to your party, but my tender heart is really quite ill' was Marie-Gabrielle's unexpected response. 'I doubt we will be able to be there. It's a long drive down and too tiring for him. Everything exhausts him these days.'

I knew that, for over a year, she had been under a great deal of stress, since she had begun nursing her ailing mother who lived somewhere in the Var, closer to us and the coast than to their remote, high-altitude quarters. So at first I understood her to be telling me that her heart had weakened under the strain.

'How is your mother?' I enquired tentatively.

'Oh, she plods along. She's as hardy as I am,' laughed Marie-Gabrielle.

It was only as the conversation developed that it dawned upon me that she was talking about François. When they were in our company, it had always tickled me to hear her call him, '*mon cœur*', or '*mon tendre cœur*', 'my heart' or 'my tender heart'. What she was now relating to me was that her beloved François was unwell. It seemed that he had suffered a breakdown. He was experiencing an acute form of grief over the loss of his 'girls'.

I tried hard to remember, shuffling through past conversations, to recall whether he had any daughters. Yes, I thought he had, from a first unsuccessful marriage to an African lady, but I was not certain of these facts. It simply did not occur to me that Marie-Gabrielle was speaking about the bees. It was only when she said, 'all his life he had dreamed of keeping hives. The loss has hit him deeply. Well, it has hit us both, of course, because we have forfeited everything; all our savings, all our investments. We have nothing but our pensions now. Even once we have managed to sell our apiary equipment, the hives, extractor and so forth, we will be obliged to convert that space and rent it out as a *gîte*.'

François and Marie-Gabrielle had been the proud owners of more than one hundred hives. The fourteen that wintered with us, housing 280,000 bees that shared our grounds, pollinating our flowers as well as those of our neighbours and environs, were but a small percentage of their stock.

'We have three hives left and they are here with us and that is it. I have put my foot down and told him, no more. We cannot go through this again. It is both emotionally and financially bankrupting. But for François, life without all those bees, "his little girls", is a lonely and inconsolable experience. Even I had not understood how profoundly such a collapse would hit him.'

'What has happened?' I asked, dreading the answers for I feared I might know the worst already.

In the interstice of time before she offered her explanation, a memory swept back. It had been a sunny early March afternoon a few years earlier when I had been alone at the farm and our apiarists had paid a visit, to see me, of course, but more importantly to confirm that their bee stock was wintering well, that the swarms had plenty to eat and were in fine fettle. After a cup of tea together on the upper terrace in the milky, early season sunshine, I strolled down with them to the lower citrus grove to assist with the uncapping of the hives. Marie-Gabrielle lit the aromatic smoke gun, the *enfumoir*, used to calm the colonies because, if they are agitated, they can sting or become aggressive, and we carefully unhooked the lid and lifted it off the first home. All was well within. Here was a very healthy and active swarm and there still remained honey for them to feed themselves on. No concerns there. The neighbouring hive presented us with a similar scenario, but then we came to the third. François was alerted even before I had removed the metal roof. His keen apiarist's instincts sensed that all was too still, too silent. We gathered as one and waited. Slowly, gingerly, I unhooked and revealed. Standing together we were, a trio of silenced onlookers. His gut feeling had been spot-on. Here was a different tale, a bleaker scenario. This entire colony of *Apis mellifera*, the Latin name meaning honey-carrying bees, lay still and lifeless. Sixty thousand lost lives.

'Such a carpet of death,' murmured Marie-Gabrielle, stroking her husband gently on the shoulder.

My colleagues spoke barely another word, but moved on in a businesslike fashion to visit the remaining families. Three out of the fourteen hives were gone, every bee dead. I was devastated. I felt that in some way we had been responsible, that some element or presence on our land had undone them, but François shook his head.

'Make us some more tea, Carol, please, and I will explain.'

That afternoon, as spring was beginning to unfold upon the world, I learned and witnessed the first details, just a sketch of the facts that, collected together, were today beginning to read like a gruesome science-fiction story.

'There is a mite. He is known as Varroa and he feeds off honeybee larvae and pupae and the damage he causes can decimate colonies. He has been on the scene as far as we know since the early 1960s but a healthy colony can usually withstand his intrusion, can fight back. It will be damaged but it is resistant and, more frequently than not, it will rally.'

'Is that what has killed off these hives?' I begged, hoping that it was and that we had not in some way destroyed or poisoned the 'girls'.

François shook his head. I saw then how upset he was though he was struggling hard to contain his emotions. A glossy black bee, quite large, that I had mistakenly taken for a small hornet, alighted on one of the flowering racemes on our magnificent wisteria climbing heavenwards alongside the front verandah. In fact, it was a Blue Carpenter bee. François leapt to his feet to point it out to me and tell me a little about it. 'It's a solitary species. They are vital pollinators for certain flowers such as that *Passiflora edulis* you have growing over there, Carol, but these little girls can also be robbers,' he was explaining excitedly. 'These Blue Carpenters are capable of slitting open the sides of flowers and stealing the pollen from within, as though emptying a safe box. It really is a neat trick! You must try and observe it. They make their homes in dead logs and dried-out, perished tree trunks. Watch out for them in your wood store, but don't be alarmed. They are not aggressive, they seldom sting.'

I was not aware of ever having spotted one before. 'Are they rare?' I asked.

'There has been a decline in their populations, too.' François ran his hands through his thinning hair and sat down again, sinking into himself.

'We have friends, fellow beekeepers who have already lost all their hives and been driven out of business altogether, but, until now, these unfortunates have tended to be living in the Var. Others over in the south-west towards the Pyrenees have also experienced traumatic results.'

'But why have they lost their hives?'

'There is an insecticide used on sunflowers, Gaucho is its name, that many believe is the culprit, but we are having difficulties proving it. And we do not have the financial resources to pit ourselves against the giant chemical companies, with their armies of lawyers.

'A healthy honeybee has an innate sense of direction and will always find her way back to her own hive. However, this product damages the bee's nervous system. She becomes stunned and is unable to locate her home. Turning in circles, disarranged, dislocated, confused, she eventually dies of exhaustion, never returning to the hive. In my opinion, this behaviour pattern has most certainly been caused by an external poison.'

Still on the telephone now, several years later, a tad wiser perhaps after my travels and a little more knowledgeable on the subject, known as CCD or Colony Collapse Disorder, I listened as Marie-Gabrielle confirmed that they had lost all their hives, save the three they were now keeping in their chalet garden. A total loss of ninety-eight previously healthy hives. And they were fairly certain, although they had not proved it, that the bees had been destroyed by legally approved chemical products used by agriculturalists on crops.

'I have already explained all this to Michel and he understands our situation. François is incapable of discussing it at present. He is too depressed. But I have refused to allow us to borrow any more money to buy new bees. They will simply be killed off. So, we will not be bringing hives to winter with you any more, Carol, though I hope you will want to stay in touch with us. We have thoroughly enjoyed our little arrangement.'

'But, of course,' I blurted, 'we count you as friends and if you possibly can I would love it if you could make it next weekend for the festivities. Otherwise, we'll find a day to drive up into the mountains and visit you both. Is there anything being done about this crisis?'

'As recently as last June, an emergency motion in the European Parliament called for certain pesticides, the neonicotinoids, to be banned in Europe until they better understand the role they play in the deaths or disappearance of honeybees. We are all hoping something will come of that.'

Bee fossils discovered in amber have been dated to forty-five million years old. Feral bees have flown this planet, foraging for their food and pollinating plants, since long before man was even a whispered thought. We know they were producing honey during the Cretaceous period (approximately one hundred million years ago) when, it is believed, flowering plants first appeared. The art of apiculture was practised in Egypt and Greece before man knew how to write, before, as far as we know, we had alphabets. In the long, slow journey of evolution, the role that *Apis mellifera* has played through its ability to pollinate has changed the physical structure of plant life. Bees are one of the earliest, if not the earliest, of earth's pollinators and, without them, the earth and all who inhabit it will be in grave trouble.

When I replaced the receiver and came to my senses, I remembered that the men, Michel and Quashia, had two weeks previously sprayed the olive trees with insecticide. The second showering of the season was due to take place the week after the party. I knew that the product they were using, though recommended to olive farmers by the Chambre d'Agriculture, was toxic. Its precise effect on bees I did not know. But as a responsible farmer, I should have taken the trouble to find out.

In the past, for one or two of our olive-growing summers, we had, at my fervent behest, relinquished the use of chemicals and had left the olives to develop naturally, without intervention, allowing Nature to take its course. Unfortunately, the results had been disastrous and we had lost the crops entirely. The fact of the matter was that no efficient, organic alternative had been found to counter the damage to the drupes caused by *Bactrocera (Dacus) oleae*, the olive fly. Dousing the trees with pesticides remained the only effective and proven method available to us.

Olive groves, especially those situated within hot, humid coastal regions such as ours, are particularly susceptible to the olive fly. It lays its eggs within the olive and the growing larva feeds off the fruit's pulp until it eventually destroys it.

The insecticide that had purportedly been causing the mass destruction of honeybees was not the same product as that used on olive crops, but the fact of the matter was that we were still engaged in the business of chemical use, covering our trees' canopies with a poison aimed not at bees, but flies, but a poison nonetheless. Our four or five sprays a year were, whether to a minor or greater degree, contributing to the problems that the planet was facing. I, who had logged my own maps of the olive's heritage and had looked full into the face of the future and seen the warning signs, was a contributor to this destruction. How could I audaciously put pen to paper if I was ignoring my own hand in the game?

It was Tuesday. The arrival of guests was imminent and we seemed far from prepared. On top of which, the pool was beginning to turn a little green; the water was crystal clear but the corners, the walls were tainted . . . It was topping thirty-five degrees in the noonday sun. Michel had cleaned it twice over the weekend, but the algae simply returned. Jacques, our handsome swimming-pool magician, a man I had counted as friend, had slipped off the radar screen. 'Not been by for weeks,' according to Quashia. Four, even five, emails, had raised no response and his phone was permanently switched to its answering machine. We had invited him to the party but even to that he had not replied. I sent one last message: We need you!

Silence. Jacques had disappeared into thin air.

I was now occupied with armfuls of plants because we had decided to adorn the house with flowers growing on the land. Quite out of the blue, as I was hurrying through the upstairs hallway, in a frenzied frame of mind – beyond the open doors, a never-ending succession of delivery vans up and down the drive – dragging swags of eucalyptus branches clustered with tiny pink and dusty-green buds that I had salvaged from an overzealous and illicit pruning by Quashia, I was halted by the ringing telephone. I picked up the receiver, hoping it might be Jacques, and jammed it under my chin while still moving on through the cool open-plan rooms, tiny blossoms falling about my feet.

'Madame, I hope you and your husband are keeping well?'

It was our *notaire*, the notary who had handled the purchase of the house and, later, the land that we had been unable to afford at the outset.

'Yes, fine, thank you . . .'

'I have phoned to inform you that I will be sending an expert to inspect your house and grounds.'

'What on earth for?'

'It is the law. We must register any irregularities.'

'What sort of irregularities are you referring to, *Maître*?' We were all pressed for time and this unexpected intrusion was not what we needed at present.

'Have you ever had an inspection for termites?'

'No, but we don't . . .'

'How about subsidence?'

Again, I responded with a negative, while scrabbling for my tallest vase. Someone was hooting impatiently beyond the window. I placed the vase on the draining board and hurried through to my den to take a look, wondering where everyone else had got to.

'And what about asbestos? Yours is an old property. I fear you are living in a habitation where there could well be asbestos linings in the roof. I see by your records that it has never been checked.'

I was baffled by this uncalled-for concern. 'But our house is built of stone, built out of the limestone rock upon which it stands, and it has a flat roof. There is no asbestos here. I wonder, could we deal with this matter next week? I don't wish to sound rude, but we are—'

'It can be hidden in the most unexpected corners. I will send someone.'

'Excuse me, is this essential?' I protested impatiently. 'I mean, it is not as though we are planning to sell the property.'

'Please take down the following name and number. Telephone the expert and make an appointment for the coming week.'

I sighed, rooting beneath newspapers for pencil and paper. I would do as I had been bidden, before it slipped my mind, but first I had to deal with the squat fellow waving a hand from out of the window of his white Renault van and after I had put the eucalyptus boughs in water. Where was everybody?

The driver was refusing to get out of the car until I chained up the dogs. 'They're quite harmless when we're here,' I assured him.

'I don't give a damn. Three Alsatians! Man, give me a break! Chain them or I'll be taking this load back with me.'

'What is it?'

'One hundred wooden fold-up chairs.'

'Did we order these?'

'Someone has.'

'Righto.' I called off the dogs which, at my approach, had begun circling the van more out of curiosity than malevolence. They came panting obediently towards me. Until my arrival on the scene, Homer and Cleo, the youngsters, had been lying in the shade right by their chains and were puzzled when I returned them to the very same spot.

'*Voilà!* Attached.'

The little man in shorts and snazzy beach shirt stepped out of his vehicle and was now grinding his cigarette stub on to our courtyard tiles.

'I've got papers to be signed,' he snapped as a lorry came paddling up towards us, belching black smoke. I could hear the crack of branches as it made its laborious approach.

'Mind the trees, the olives, the young figs! Oh, for heaven's sake!'

An elongated blue cube, rather like a telephone kiosk, was the master of the destruction. What on earth was it? Where *was* everybody?

'Lady, do you want these chairs or not? I've got another delivery to make before lunch. If so, sign the invoice and I'll be back for them next Monday, all right?'

'Yes, yes, of course.' My concentration was still with the kiosk. I signed without reading the paper and bid the chap *bonne journée*.

'I've got to unload them yet. Where do you want them?'

'Erm, anywhere,' I sighed. He took me at my word and unpacked them smack bang in the centre of the parking area, then, swerving by the old lorry, shot off down the drive. The pile of chairs would now hinder the newly arrived's exit when it was ready to go. In fact, until they had been restacked elsewhere, they would hinder all traffic. I began to shift them two by two, stacking them up against the garage wall.

A bloke with pencil-thin moustache and round fat face beaded with sweat poked his head from out of the cab.

'Oi, lady! Are those Alsatians attached? Cos, if they're not, I'm out of here. What a sodding entrance. No one mentioned that the drop-off would be as difficult as this. It would've cost extra. Anyway, I'm 'ere now. So, are they chained?'

'Yes, they are.' I was wilting. 'Could you tell me what you have, please?'

'Portaloo, mate. Where do you want it? I'll install it, like it says in the contract, but it has to be on flat land or it leaks. That's what was signed and quoted for and I'm not carrying it anywhere over those bleedin' terraces.'

'No, of course not.'

'So, where do you want it?'

It was big and ugly. I did not want it at all and could not imagine why it had been ordered. 'If I reverse that old VW, could we position it in there, hide it behind those hanging fig branches, so that it's out of sight?'

'Anywhere you want, sweetheart. Which way do you want the door then?'

I had no idea. I inched our farm van back, avoiding the chairs and the disgruntled dogs, and once the pulley system had landed the lavatory on to the tarmac and negotiated it behind the fig with little collateral damage we settled on the direction for entering and exiting this undesired latrine. I could not think who would be using it.

'Sign here, if you will. I'll be back for it on Monday, as agreed. Any chance of a beer, love? It's bleedin' hot. Nice place you got here.'

I scribbled where his podgy finger was pointing. 'I'll just get the beer, wait a sec.'

Within the cool darkness of the garage, still reeking of wine as well as diesel and oil seeping from the various gardening machines housed there, I found another half a dozen fragments of ceiling plaster splattered on the ground. Stepping over them, I reached for the beer, took it to the driver who was now over in the shade talking to Cleopatra, the most approachable of the dogs.

'I love dogs,' he grinned. 'Cheers, good health!'

When I returned inside the house, the telephone was ringing. I was tempted to let it go to answering machine but was still holding out hope of news from Jacques.

'Allo?'

A woman at the other end: the expert recommended by our notary. The call I was supposed to make had slipped my mind. She was suggesting a visit in two days' time, warning me that the inspection would take the best part of a day and that she would require photocopies of all land plans and contracts as well as several other documents, all of which had been filed somewhere, but I was not entirely sure where.

'I wonder, might we, please, leave this till next week. We're having a par—'

'I'm fully booked next week. I have spoken to *Maître* —— who informed me that the matter is urgent. You should have telephoned me. Never mind. I have set Thursday aside for you and I will be there by ten. Please have the paperwork ready. *Bonne journée, Madame.*'

That evening, the first of the pre-party guests arrived, an actor friend of mine from London whom I had not seen for almost two decades. He could not be with us for the big event on Saturday so he had arranged with Michel to spend a few days with us in advance. He was willing to lend a hand, but we would also have to understand that he was giving a poetry recital on the following Sunday and would need to learn his lines and rehearse them.

I had also just learned from Michel that neither of his daughters, twin girls, Vanessa and Clarisse, would be with us. Both were mothers now and their commitments lay with their offspring. Travelling with a troop of toddlers was expensive, logistically complicated and, in any case, our modest farmhouse offered limited facilities for these burgeoning families. I was rather disappointed. It had been some time since I had seen them. The delightful news was that Clarisse was pregnant with her second child. A first-week-of-December *accouchement* was expected. 'So, by Christmas,' I winked at Michel, 'you'll be five times a grandfather. Where does the time go?'

'Perhaps we can organise a gathering at Christmas. All of us together,' he said. 'I'd like that.'

'Shall we plan for it here, a big family Christmas?'

'We'd be bursting at the seams. At least this weekend all those who are not in hotels can sleep in tents. Hardly possible in wintertime,' he laughed.

Michel saw his girls so rarely, due to overloaded schedules on everyone's part, and I knew he had also been looking forward to sharing this party with them. It was not for the first time that I regretted our farmhouse counted only three functioning bedrooms – years earlier Michel had removed most of the interior walls on the upper level – plus a fourth that had long since been requisitioned by me for use as a den, a working space. When needs must, a mattress was laid out on its floor and it metamorphosed back into temporary sleeping quarters, but this was not an arrangement conducive to the presence of small children. The fact was that our house was inadequate for the size of the fold we now represented. It was impractical, I was fully aware of that, and it had been far too long since the girls – we still referred to them as *les filles* in spite of the fact that both were mothers now – had come to stay with us.

We sat in the sweetly scented shade beneath the flowering

Magnolia grandiflora, Bridget (who had ordered the Portaloo), along with Michel and our friends from Berlin, Hans and Sabine. They had spent the better part of the day traipsing about in the brittle heat of Cannes choosing and ordering the 'birthday' cake. It was to be a surprise and they displayed much jollity and secrecy about it. I was uncorking rosé. Roger who had been showering and had slipped into linen shorts and shirt and panama, the Englishman abroad, joined us.

Those tranquil moments of aperitif were the last we enjoyed for days. While the men prepared the barbecue – we had three installed now as Michel had insisted that he was going to organise all the catering – 'we are not going to call in a firm and pay those exorbitant prices' – we girls picked herbs, uncorked bottles and laid the table. I could not imagine how Michel thought he would handle such a culinary undertaking. We had estimated that between late Friday afternoon through to the following Monday morning, as the party numbers waxed and waned, we would be requiring, for the three evening meals alone (to say nothing of the lunches, brunches, breakfasts, snacks, aperitifs), two hundred and thirty-eight covers. It was so far-fetched that I had stopped worrying about it all (except for the moments when I hadn't). Beneath the soup-plate sized, white waxy magnolia flowers now in full blossom, each of us was allocated our tasks for the following three days.

I was collecting my mother, a hearty eighty-four years old now, from the airport. Many were flying in and hiring cars. We had guests coming from as far afield as Los Angeles though Michel warned that there were bound to be last-minute cancellations. Bridget was lending us a mattress or two while her partner, Luigi, who was a local *boules* champion, was going to be organising Saturday and Sunday afternoon tournaments on one of the terraces.

I was despatched to some ghastly hypermarket to buy five new dustbins, two dozen rolls of biodegradable dustbin bags, one hundred knives, forks, plates, what felt like a hundredweight of candles, half a ton of pistachio nuts . . . the list was endless.

'What have we forgotten?'

Replacements for the smashed wine bottles and replenishment of the booze already being consumed daily.

'Have we forgotten anything else? Anybody, any ideas?'

The men were slinging sails higher up the land between a quartet of Italian cypresses. 'We need more rope. Rigging rope and metal hooks.'

I had come up with the idea of creating in the garden (in the loosest sense of the word) a pleasance, which, in recognition of my travels, would be known as the Bedouin Bar. The corner we had chosen was discreet, suitably distanced from the main house, and looked out over both the Mediterranean and, to the right, the mountains, while the festooned sails sheltered it from the harsh sunlight. It was ideal, a perfect spot for nattering, congregating, reading and, of an early evening, sipping glasses of chilled wine. Once dressed, it would require a fridge, low table, sofas, garden glasses, cups, cushions, coffee machine . . . phew! Quashia was shoring up the drystone wall to its rear, to provide its seating area with a certain rough-hewn definition.

Poor Quashia, he was completely bemused by the level of activity.

And the phones never stopped ringing. 'Darling, we've arrived!'

'We're lying by the pool at the hotel. Gorgeous weather. So dreary back home.'

'Longing to see you. Anything we can do? No, well, we'll see you Friday evening then for the kick-off. What's the dress code?'

'Darling, I am so sorry. I can't be there, after all. I'm night-shooting Friday and . . .'

And . . . and . . .

'Hello, we're lost! We hired a car and took the motorway towards . . .'

Roger was prancing about the land with a script in his hand, declaiming, concerned that his hat might fail as sufficient protection against the sun's damaging UV rays. 'I don't want to burn my face, darling,' he repeated to me on several occasions. 'I've got a show on Sunday, don't you know. Such a pity I can't stay for the bash. What creams have you got?'

Whenever I entered the upstairs area of the house, there I'd find him, rooting about in the bathroom cupboards in search of moisturiser. Or at my computer on the internet emailing his agent.

New friends, old friends, long-lost colleagues: the grounds began to buzz with them. Ipod programmed to shuffle and portable SoundDock gave us non-stop music: reggae, romantic, dance, jazz, cool, eclectic. My mother, who remained hale and hearty but was getting deafer and suffered a tremor as a result of the many trials she had endured since the death of my father, was ever at the thick of it all, opening bottles for our invited, slathering herself in sun cream (instantly pinched from her by Roger), dressed elegantly, shouting to be heard, assuming the whole world was as deaf as she was. By breakfast on Thursday morning our in-house numbers had reached ten, so house bursting as well as three tents rigged, and many more to come. Not bad going for what was a mere three-bedroom property. Roger was leaving the following morning and his room would be taken by my ebullient agent and his wife. I was constantly hauling out sheets, washing towels, counting out bowls, mugs, glasses, checking lists, while Michel and Hans returned every three hours from the shops and farmers' markets weighed down with produce. I was delighted to hear that Michel had finally decided against roasting a lamb on a spit up near the Bedouin Bar, which was now dressed and was sinking-into-deep-cushions appealing.

'No spit, please. Too hot, too risky. We'll have the fire brigade after us.'

The forecast was threatening a storm, but the sky loured stubbornly and the rains never came. So busy was I with such inconsequentials, any one of which might have put a spanner in the works, that I had completely forgotten it was Thursday morning. A rude awakening when I spied a royal blue Noddy car zooming up the drive.

Had I overlooked the arrival of someone? Miscalculated? Were not the next assembly of guests due the following morning? The vehicle drew to a sharp halt and out stepped Madame l'Expert.

La Madame, probably in her late fifties with severely shorn iron-grey hair and narrow glasses, had arrived promptly, unusual in this area, at 10 a.m. two days after she had telephoned, acting upon instructions from our *notaire*, she confirmed. She was wielding a briefcase, handbag and magnifying glass and was wearing a tight skirt and heeled shoes. 'Right, let's get to work. What have you constructed since you moved in here and what is original?'

I pointed out our humble additions, including Quashia's *hangar* and our rather splendid greenhouse – 'all glass and steel, no asbestos there,' I assured her with a smile.

She raised her eyebrows. 'You have no idea, Madame, where it can be lurking. Asbestos can be found in three thousand products in the home and industry. Pipe insulation, heating ducts, for example. In a property of this age, I doubt I will go away empty-handed.'

The slanted roof of our *hangar* was tiled, a Provençal style known as *tuile*, from the Old French word *teuille*, or tile, and originating from the Latin *tegula*. These tiles are fired from earth, terracotta, and are long, curved and sit one on top of the tip of the other. Frequently, they are laid on a corrugated bed of fibre-cement sheets, which was the case here. From the parking area she scanned the hillside, and settled her attention on the *hangar*. *Amiante*, she hissed under her breath. 'I will begin with the *hangar* and the greenhouse.' She commenced her ascent, up the steps, striding resolutely, an arrow to target as she soared by Roger whose arms were outstretched, papers flapping in one hand, reciting sonnets of the Bard, all without a glance in his direction. Shoes and skirt no hindrance to her attack on the potential asbestos. The dogs were darting about at her feet but they did not trouble her either. She marched to and fro beneath the roof and then shook her head. 'When did you say this was constructed?'

'About four years ago, I think.'

'That explains why it is not asbestos. The product was taken off the market long before.'

I glanced across at the others hard at work, knowing that they were waiting for me to get to the shops for the requested bits and bobs of hardware.

'I really think you are wasting your time,' I implored.

She spun on her heels and bored a look into me that might have nailed me to the wall. 'Do you realise that asbestos is a toxic material? Not only is it dangerous but it can be carcinogenic, which is why my work is vital. *Vital*. I will find it, identify it, list it, and file it with the cadastral papers of this site, and there it stays in your house records until it has been dealt with. You realise, too, Madame, that, should I detect asbestos here, you will not be legally entitled to sell this property until you have replaced it.'

This, I learned later, was inaccurate, an exaggeration. One was obliged by law to register and declare asbestos in order that a potential purchaser was fully aware of what he was taking on and not going into the act unwittingly, but the house could change hands.

'I accept what you say, Madame. However, I am fairly certain that no asbestos exists here. And, if I may say so, I am also rather amazed that the same rigorous attention is not being given to the insecticides sold in shopping centres and supermarkets, products that any part-time gardener can spray willy-nilly, where he likes, without redress. Pesticides that are also dangerous to one's health.'

Those black eyes studied me uncertainly. Was I mocking her and her vocation? 'I know nothing about gardening. It is not my expertise and I hate it.'

'But you are in the business of tracking down toxic materials . . .'

'Not at all! I deal exclusively with *amiante*, asbestos, but let us get on with it or I will be obliged to return tomorrow and the fee will be double.'

We traipsed the land, up and down the hillside, Bruce Springsteen reverberating round the treetops. Every water basin, tap and pipe surround was examined and found to be in accordance with the laws. Nothing was amiss. I was beginning to feel rather proud, personally responsible for our residence's exceptionally clean bill of health. By now, we had been at it for several hours. The rest of the troop had stopped for refreshments, followed by lunch. I needed a swim and suggested we pause for a cold drink, which this lady accepted. I discovered that she was recently widowed, that her husband had been an esteemed local architect and that, since his death, because she had done all the inspections on the sites where he had been building or reconstructing, business had been less than brisk.

Was this the reason the notary, a personal friend of her deceased husband, I also gathered, had telephoned me? Had he been drumming up trade for the woman who was down on her luck? But I had softened and felt rather sympathetic towards her. She roamed the interior of the house, craning her head, squinting into corners, picking up and examining objects from my travels and seemed genuinely interested and taken with our motley collection of furnishings and *objets d'art*. I talked a little about my journey circumnavigating the

Med and my passion for all things olive, while she sipped her lemonade and listened, watching me, scrutinising me as though I too might be a harbinger of toxicity.

'Olives hold no fascination for me,' was her bald response. But afterwards she smiled and said, 'You are fine. There is no asbestos here. I don't think I need to look further. Well, I'll just do a swift recce in the garage and then I'll be on my way.' She was marking sheets of paper with a tight, frenzied energy, sheet after sheet. Each page contained lists of questions, descriptions. In every instance, she ticked the boxes.

'Positively in keeping with the law,' cracking a rictus.

We strolled out into the sunshine, making our way down to the parking, alongside the garage. In the distance, Van Morrison's *Astral Weeks* could be heard.

'Are you in the habit of hosting such a full house?' she asked.

I shrugged.

'Do you mind it?'

'I enjoy it once in a while. We're preparing for a party.' I threw a glance towards the others who seemed to be having a very jolly day, singing along to a seventies tune I recognised but could not identify. I smelt fresh paint and wondered what was being decorated. Elsewhere, someone had a barbecue on the go, herby sausages were being grilled. A cork was popped.

'Since my husband died, I find time rather a disconsolate, demanding companion. Is the garage open or do you have the key?'

'No, it's unlocked but it's in an appalling mess, jam-packed with junk and machinery.' I was embarrassed by its condition. 'And it reeks of damp and er – of wine, due to the leaks in its roof. There are tiles, clods of plaster falling regularly, so please watch your step. You'll have to climb over the debris, I'm afraid, or wait while my gardener and I clear it out.'

'Not at all, I can step.' She raised one leg as far as her skirt would allow and pointed her foot like a ballerina. Suddenly, she was a girl again.

'Let me come in with you and put on the light. At the rear to the left is the heating room. It's a boiler from the dark ages but, until we can afford to change over to solar heating, the dinosaur works and we prefer to leave it as it is. The boiler room was tunnelled out of the rock upon which the rear of the house sits so I doubt you'll find anything toxic in there.'

She sniffed about in the crepuscular light. 'What a shocking condition this garage is in,' she was bending forward, arching like a switchblade, flicking on and off a delicate, pencil-thin hand torch. 'You should get it seen to before it collapses entirely.'

'Yes, well, we intend to.'

Then rising to her full height, she declared that she was satisfied.

We had passed muster.

'Well, that's a relief,' I grinned.

'Here's my card. I can recommend you to the architectural firm my late husband founded. They'll fix that ceiling for you in no time.'

Out we went again and strolled to her little royal blue car. It reminded me of a jelly bean. Still talking, she was unlocking the door, before tossing her case and papers on to the passenger seat and then she swung back round to shake my hand. As she did so, something behind me caught her attention. Glancing about, eyes darting this way and that, her face began to convulse. It was as though she had set eyes on a monster.

'Asbestos!' she cried triumphantly. 'How could I not have spotted it? Right there, under my nose. What are those two box rooms used for, stables?'

'Exactly. The dogs sleep in one and there's our washing machine and junk for the party . . .'

She was not listening. Swiftly, out came a tape measure,

pocket digital camera and a small serrated knife. From her handbag she withdrew a white mask. This she tugged over her head to cover her nose and mouth, suddenly resembling a dentist or surgeon.

'Do you have a ladder?' she drawled from behind the mask.

'Sorry?'

'Ladder!'

'Oh, it's right beside the stable doors. There, see it?' I responded, 'but I don't think that roof is asbestos. Those are corrugated sheets . . .'

She pulled on the elastic of her mask and lifted it clear of her features so that she could be understood. 'And they are made from *asbestos*. *Amiante*, right here in front of me.' She was up on the ladder, digging furiously at a small section of the roof's overhang. Her manner had become evangelical, demoniac. 'I know I'm right. Nonetheless, it will have to be sent away to a laboratory for testing and confirmation. Oh, this is bad news. Very bad.'

She sounded positively joyful.

'Well, how bad? We can remove it. My husband and the gardener, if necessary with my assistance, we can have those sheets off there in three-quarters of an hour and we can retile it as we have done the *hangar*. It's no problem.'

'But it is. Don't you dare to touch it! If you do, you will be committing an offence. And then you will never sell this property. I must speak to the notary and inform him.'

The triangle of flecks that she had by now carved off the front of the roof were being wrapped in plastic and then sealed in a tin and packed once more in an outer wall of plastic. This was then placed in the boot of her Clio.

'I am afraid you will be hearing from me. I am also sorry to say that this is going to be very much more expensive than I had originally anticipated, costed, because there will be extra notes to write up as well as the laboratory fees and their report. Good afternoon.' And with that she revved up the engine and gunned off down the drive, a Valkyrie on a mission.

Of the five houses occupying our hill, two skirted its hem, one other stood guard at the entrance to the lane, the fourth had been built behind us on the darker, inland-facing slope, while we occupied the south-facing, sea-view side. We barely, if ever, encountered the proprietors at the rear. In fact, they could have knocked their house down and we would have been none the wiser. The neighbours at the entrance to the lane, we ran across very occasionally. I knew when they were in residence because I saw their dustbins. That was about the extent of our intimacy. The property that lay to the west of ours had suffered a series of unlikely tragedies. Each of the proprietors, three in reasonably quick succession, had died, having fallen foul to a fatal heart attack in their sleep. It struck me as most bizarre. Three in a row, victims of the same fate. What was more, they were all Italians. That house now stood empty. The remaining pair of properties were, to a lesser or greater degree, a part of our lives here. Both of these had been sold during my absence and we had yet to meet the new occupiers.

It is etiquette in France that when someone moves into the street you send a little note, to introduce yourself and invite the new arrivals for an aperitif. Due to all my absences and to Michel's own overcharged programme, we had never fulfilled this duty and regretfully the last of the Italians had died before we ever set eyes on him. Michel and I agreed that the weekend bash was the perfect moment to redress our oversight and invite the neighbours all along for a drink.

I dug out some cards in my den and wrote a little note to each of them. 'Please enjoy a drink with us during our upcoming festivities, any time that suits you over the weekend' and apologising in advance, forewarning them of the extra activity in the lane, cars parked across on the grassy bank next to our cottage, the music. General excuses for all inconveniences.

The following day, Friday, I was walking the lane, delivering the envelopes, when the Moroccan lady who occupied the most modest of the properties, right alongside our cottage where Quashia resided, drew her car up alongside me and wound down her window.

'Good morning,' I smiled. 'We haven't met yet. I'm the lady from the house on the hill, the olive farm. I have just left a note in your box. We'll be having a party and I wanted to apologise for any—'

'If you don't keep your dogs under control, we'll call the police and have them put down.'

'What have they done?' I stammered.

'Breach of the peace,' she snapped and shot away.

It was now Saturday and the skies had stayed clear. The van that had delivered the folding chairs had returned with nine, circular fold-up wooden tables. I signed for them, wished the driver '*bon weekend*' and glowered at him as he ground another cigarette underfoot on to recently swept tiles. The washing machines were churning, the dishwashers whirring. We had been thirty-eight beneath a starlit sky for dinner the previous evening and the latest calculation for the big evening was that we would be eighty-seven. As Michel had forecast, there had been several cancellations but there had also been an addition or two. Marie-Gabrielle had telephoned to say that she and François would make an appearance at some point. How could they miss out on such a welcome home party? Six had appeared from Malta, olive farmers Nat and Julia, a firm friendship that had been bonded while I had been on the island during my travels. With them came journalists, friends of theirs who had also become friends of mine. They came bearing gifts in abundance: perfume, candles, fish, squid and copious amounts of wine plus a bottle, of course, of their olive oil. Nat was proposing to prepare a cuttlefish risotto for eighty! With his characteristic energy, as soon as they had arrived and offloaded bags, shawls, pashminas, he had set himself up a chopping board at a table beneath our jacaranda tree and set to work. The others were socialising, chattering in their native tongues or in Italian, a little French and lyrical English. A young olive farmer drove in from Sicily. He had departed his south-western spot some two weeks earlier and had taken his time, moseying north with his wife whom I had never met before. I did not know then, only months and months later, that he had been diagnosed with a terminal cancer.

Others hailed from Manchester, also new friends I had met elsewhere on my travels. The sight of chums from drama school whose faces remained glamorous, elegant, but were now lined and matured, reduced me to tears. Michel was acquainting himself with men and women I had never even spoken of because they represented a past that belonged to another era, to Carol, the actress and single woman. When the beekeepers appeared I greeted them with hugs but was profoundly shocked to see the frail man, led by his wife, who had once been François. Publishing colleagues, two doctors, olive farmers, filmmakers, actors, beekeepers, sound technicians – who would have thought they would have found any worlds in common, but they did. How had Michel put all this together? A few children accompanied their parents. Our trio of canines were sporting brightly coloured kerchiefs, variations of red, and paid not the blindest bit of attention to most of the comings and goings. Bridget, who was supposed to have been organising the ice, phoned in sick at the last moment and there was a post-midday panic when we realised that there was nothing with which to keep the gallons of wine and beer cool because every fridge was clogged with food. Eventually, it was dealt with along with every other minor hiccup by one guest or another. Everyone lent a hand. The *boules* tournament was an even greater success than I had dared to hope. Throughout the long hot afternoon, a punctuation to the relentless cicadas, the popping of corks, was that soft click of metal ball against metal ball, followed by the floating cheers and cries, '*Bien joué*', 'Well played, *mon ami*'. When I paused, took a moment from the serving of drinks, the unwrapping of gifts, the hugs and the introductions, I tuned in to voices coming from everywhere across the land and I lost count of the number of languages being spoken. Two Chinese brothers were boiling rice in the kitchen, a friend from South Africa, Bill, along with Michel, Hans and others struck the matches to ignite the barbecues. Meats, grilled fishes and vegetables were sizzling . . .

Marion, my agent's wife, took it upon herself as evening drew in and the pool was emptying, sky-blue reflected in the puddles of splashed footprints everywhere, while folk were showering and donning their glad rags, to gather together my host of candles and exotic lamps from lands afar – Moroccan lanterns, a tall Lebanese hand-beaten copper one, others transported from Tunisia, to decorate the garden, to hang Vietnamese paper lanterns from the trees.

'Go and dress for your own party,' Marion whispered to me. 'You are the one who has come home.'

When I returned outside, she had created a wonderland. I had never seen the terraces so enchanted.

The highlight came after the meal, which in itself was an occasion never to be forgotten. Bridget, now alongside her beloved Luigi, had finally arrived bearing the cake moments before dinner and had secreted it somewhere cool and safe. It was wheeled out with champagne as a waxing moon was riding high in the sky. And what a cry of joy when I set eyes on it; it had been fashioned in the shape of an olive tree.

After a speech that had brought a tear to almost every eye, Michel led the toasts, 'Welcome home to my wife. I am very glad that she has returned safely.'

Someone, I cannot remember who, put an arm across my shoulder and whispered: 'You really have created something special here. What could possibly destroy such a paradise?'

2

Appassionata was a very modest domain set in a magical location, furnished with stones and walls and ancient trees, haunted by birds, squirrels, wild boar, lizards, geckos, dogs . . . life. A great percentage of its lands had been sold off after the Second World War and today its grounds measured a mere 29,000 square metres or, approximately, three hectares. These were made up of a wild, self-seeding pine forest that climbed at an angle up behind the house to the apex of its soft-underfoot hill, while the remainder was dominated by both venerable and young olive groves – the juniors planted by us – and a few terraces of adolescent fruit trees, also planted by us. The farmhouse, an Italianate cream villa wrapped about with verandahs and balustrades, was built in 1904 by a rich merchant family from Milan. Constructed halfway up its hillside on a base of limestone, its surrounding grounds had been modelled with drystone walled terraces: *en restanque*, in Provençal, a quintessential Mediterranean image and, as far as I was aware, the oldest form of manmade irrigation around this tideless sea.

When my lover, now my husband, Michel, and I found Appassionata and its lands, this lovely dominion had been abandoned. Its grounds had become a jungled ruin, forgotten and uncared for, but we both fell instantly in love with it and its breathtaking views overlooking the turquoise waters of the Mediterranean. To describe its purchase as rash would be an understatement. Few in their sane mind would have taken such a risk and certainly not a couple living separate lives in different European countries, recently acquainted and lacking the necessary financial resources, never mind the expertise required for such house restoration. We did not possess even the deposit, let alone the asking price, and we had no real notion of what lay beneath the unruly jungle that greeted us on our earliest visits. It was impossible to estimate the sums that would be required to make the once elegant house even basically habitable, while to reinstate its natural grace and beauty was beyond our wildest financial ambitions.

The estate agent and Belgian proprietors assured us that we were purchasing a viable farm with mature olive groves and two acres of vineyards. One of its other selling points was that it had its own private, registered water source. It was not marked on the *cadastre*, and to this day we have never uncovered that well, though Michel has scoured every square centimetre of the grounds.

Somehow we became the proprietors of the place, purchasing it in two parcels over a period of six years because we could not scratch together or borrow all that was required at the outset. Even today, we refer to half the land – the half without the house but with its own rather romantic stone ruin, that must during the first half of the last century have been a cottage, a bothy for the gardener, vine-tender or possibly goatherd – as the 'Second Plot'. Because of our penury, it took two years before we could afford to hire a gardening firm to cut back our treasured jungle, including the Second Plot – all rights to it were signed over to us at the same time as the purchase of the house even though we did not yet legally own that parcel of land – and when the weedbeaters and tractors had packed up and lumbered away from this hillside, having bared earth that had not seen the light of the sun for at least a decade, there, revealed in all their knarred and statuesque, almost metallic beauty, were sixty-eight four-hundred-year-old olive trees growing along the hill's lower terraces. There were many other marvellous discoveries, too, but none as awe-inspiring or as far-reaching for my destiny as those olive trees.

I and the man I had fallen in love with had not been looking for a farm. I am an Anglo-Irish lass, more Irish than English, and spent a great deal of my childhood on the family farm in Ireland, but that life bore no similarity to producing olive crops.

With impeccable timing, serendipity stepped in and blessed us with a retired Provençal, René, who niftily assumed the role of Appassionata's olive guru. He taught us, me in particular, how to tend olives, how to gather and press them, and by the end of that first olive-harvesting season we had produced our very own, premier-class, deliciously peppery olive oil that later went on to gain us the coveted AOC ticket, an Appellation d'Origine Contrôlée, which here, in France, is the benchmark of top-quality produce. What a bonus! Not only my dreamed-of House by the Sea and the man to share it with, but a most unexpected gift from Nature to accompany it.

Olive farming was not my trade. It still is not, but it very soon became a passion. However, alongside the physical activities of farming the fruits and pressing them into oil, I am equally if not more passionate about the tree itself and its role within Mediterranean cultures. The olive is a sacred tree and civilisations since the beginning of time have honoured it as such. It is also a very gifted and accomplished organism and one that might very well have a role to play in our planet's precarious future . . .

These revelations came as treasures, surprise investitures

secluded within the jungled heart of the property. They did not reveal themselves all at once – of course not – but little by little, one delight after another bestowed upon us. I loved this House by the Sea (as I had christened it long before I had ever found it), and was profoundly happy with my lot. I was excited, passionate about its fruits of the earth and I shared this idyll with another whom I also loved profoundly, my husband, Michel.

As the marvels, the discoveries – I am referring in particular to the fruits and, most importantly, the olives and oil production from the estate – were harvested, we celebrated them joyously. Encouraged by René, we sought better, bigger harvests, but in order to produce this exceptional oil of ours we had first to combat a very resilient predator: the olive fly. René had taught us that spraying the trees against this indomitable enemy was the only course possible and within five or six years of taking ownership of Appassionata, a wooded tract of land that had been left to its own devices for a decade, we were covering the grounds with pesticides. I objected, but there was no alternative, I was told. My own role in the future of such farming practices had not yet become clear to me.

During the days and then the soporific, heat-ridden weeks beyond my return, beyond a forever-memorable homecoming bash, while the men, my husband and Quashia, laboured and sweated, attempting unsuccessfully to repair the impoverished condition of the garage ceiling falling like hailstones around them, or were reconstructing the drystone walls destroyed by trespassing wild boar, I was netting the trunks of the apple trees. These were the principal target of the pigs' infernal nocturnal raids, when they were not nosing in the walls for snails. It was not uncommon to wake and find broken branches scattered on the grass, others snapped and hanging like damaged wings. One slender tree had been severed at its base and lay expiring on the ground. Its fruit, hard, round as golf balls, had been snaffled.

Letters arrived. The first was a detailed laboratory report stating that the corrugated roof straddling our two stables was 100 per cent asbestos. The second was from l'Expert herself informing us that these facts had now been registered with our notary and would be added to our house files. She suggested that we speak with the *notaire* directly; as well she requested that her bill be settled within the week. The invoice was included. A call to the notary confirmed that he had forwarded the results from the laboratory to the municipal offices in Juan-les-Pins which held the classified records belonging to this property.

'When you have organised the removal of the offending corrugated sheets, please make sure that you send me the original certificate of disposal, signed, dated and stamped by the firm who carry out the work.'

'Could you, please, give me the contact details of any firm who—'

'Madame, I am a *notaire*. I am not allowed to hand out recommendations. You must find someone yourself.'

'What about l'Expert, might she be allow—?'

'Her situation is the same as mine. I look forward to receiving the relevant certificate in due course. *Bonne journée.*'

The second round of insecticide spraying was upon us. I was upstairs still unpacking belongings when I heard the men unlocking the garage and I hurtled down the stairs to raise my well-played vociferous objections. I found them within the crepuscular light, gazing into air that was dank.

'Something has to be done about this place before the whole blooming side of the house caves in.' This was Quashia's gloomy forecast for the condition of our garage. There had been an overnight summer storm, violent and splendid, and the water penetration was visible, stressing and splitting the precarious hold of what little plaster remained. Added to which the humidity was damaging the tools and equipment housed within there.

'We'll have to start covering everything in plastic sheeting or find somewhere else to park it all. Come on, let's get to work.'

The issue of insecticide usage had become a bone of contention yet again, but I could offer no concrete alternatives and they were unwilling even to hear me out now. We had lost our entire crop the last time we had chosen to follow my hands-off route. Eventually, I gave up arguing and returned my thoughts to my journeys while the protection of our fruits continued, drenched by the same unsatisfactory system. On the days such as today – once every twenty-two – when the machines and pesticides were hauled out and the liquids mixed into containers and readied, I took myself off the premises, muttering that I almost wished the garage would cave in and crash down upon these obnoxious poisons that hung in the air smelling foul for hours after the work had been completed. I dared not dwell upon the risks to the wellbeing of my husband and gardener, even protected by face masks. I began to feel isolated by this dilemma. How could I, recently returned, remain on this olive farm while our methods for its fruit production troubled me so profoundly? As I could offer no sound alternative, I silently questioned whether we shouldn't simply sell up and leave our beloved Appassionata.

The fact was that I was unsettled. I was restless, bearings askew. I was physically home, but I was still everywhere, hopping between images, trying to make sense of a mass of material: photographs, notebooks, yes, and the arguably more powerful, frequently haunting, images in my head.

To counter the doubts, I took delight in the humming perfumes of the mountain-growing *garrigue*. I documented every flower and weed that passed underfoot, most of which had by now been cut back due to the fire risks, leaving the hillside a mass of felled weeds and grasses. I stared in dazed amazement at the silhouette of every tree, absorbing its form, comparing it, always recalling other trees more ancient, different, pruned in creative and curious fashions.

The history of the olive tree, its undiscovered origins, remained swathed in mystery, obscurity. I had not returned with the Holy Grail, the jewel box that contained the rarest, most precious of testimonies, tantalising clues from shadowy, unrecorded pasts. Those elusive records had evaded me, which was partially why my quest remained incomplete and why I was feeling rootless and routeless. But I had gathered together a million stories from manifold pasts and I wanted these to be the links for a new project I was growing excited about. The creation of an Olive Heritage Trail. It was to consist of voices, songs, witness statements, olive presses, farmers, scientists, museums, manuscripts, texts, photographs, footage, long-forgotten particles of stones, wheels, jars buried in fields and deserts; a broad range of links reconnecting a fast-disappearing history that embraced the Mediterranean. Sailed by galleons, trekked by caravans, walked by beasts, herders, labourers, crisscrossing waters, sands, deserts, mountains and coastlines; traces of the human passage connecting disparate elements from this small sea and the civilisations that have inhabited its rim – and every one of these clues pointed to, or claimed at its core, the olive tree. It was indeed a treasure chest and I was attempting to prise it open, to reveal its contents as a jigsaw puzzle, or many jigsaw puzzles, to be pieced together, to paint pictures, conjure up scenes, scenarios to please the viewer, the visitor of today and tomorrow; to honour the olive tree and its workers and players, to dust them down and bring them back to life, risen from their yesteryears.

All this was a long way from real life and domestic responsibilities.

Added to which, another portfolio of images, a twenty-first-century map of the Mediterranean, was painting a more troubling scenario. It was this file, these discoveries along my journeys, that was causing my inquietude, possibly overwhelming me. My quest had set course for the past, but I had also encountered the future, drawing closer by the day, threatening a bleak tomorrow. It was studded with warning signs, arrows pointing to desiccation, desertification, signalling an unproductive, unpropitious tomorrow. The Earth's Destruction. And I did not wish to be a contributing factor to that scorched earth.

By now it was cicada-rasping August: bone-dry days, the crisp perfume of pine, a state of abandoned lethargy. Guests arriving, departing, sinking themselves into the *fraîcheur* of the pool, shaded lunches that extended beyond the soporific hours of the siesta through to the next round of aperitifs. Out of the blue, a call from René. He was proposing an invitation to Raymond's homestead, the wealthy business tycoon who had turned water diviner and olive farmer. They had much to celebrate, René bragged.

This was the René who once upon a time had been my teacher, our guide in all matters concerning *oléiculture*. I had bowed to him as my olive guru until eventually his practice of using excessive quantities of chemicals had driven us apart. René sat most firmly on the side of those who sprayed without concern and happily killed off anything if he thought it would deliver him larger returns. On this issue, we had always been at loggerheads. He, like Quashia today, judged my methods and ideologies as 'naive, impractical'.

Still, we had remained firm friends, but friends who met infrequently. I was extremely fond of him and, ironically, held a rather soft spot in my heart for him and his Provençal wiliness, which, under different circumstances, I might have described as downright trickery. René was well practised in the arts of artifice.

'Raymond wants to host a lunch party in your honour. We want you to see the trees, to photograph our triumphs.'

Before the Second World War, when Raymond and René had both been little more than a pair of nippers, they had attended the same village school, not a hand's throw from our farm, and at that time they had become fast and firm buddies. Beyond those kiddie years, life had sent them off in very different directions and had never reunited them. Until five or six years before this conversation, they had not clapped eyes on one another for more than seven decades. After Raymond retired from his hugely remunerative hardware business, he decided to concentrate his energies on his untapped talents as a water sorcerer and to consider the potential for farming the trees that had lain idle on his olive farm for generations. His dominion – attractive homestead, gently sloping fields a few miles inland of us – had boasted five hundred mature *oliviers*. Raymond openly admitted that he had known nothing about any form of farming and, to obtain the best returns possible, he required expertise. That was when René fortuitously walked back into his life. The two old-timers shared between them well over one hundred and fifty years of memories, and they hit it off at once. René, when asked by Raymond his thoughts on how best to manage the groves, had cleverly advised: a) the pruning of the old masters, b) an extension of the plantation. Raymond grabbed his childhood friend's advice with enthusiasm and together they planted up another two hundred trees. These, of course, required husbanding in a way that the elders did not. Thus René had placed himself in the position of becoming indispensable to Raymond.

A short while back, a year or so after their reunion, they went into partnership together. René took over the reins of Raymond's olive farm and, it had to be admitted, he had done his bit towards metamorphosing the moribund estate into a highly respected and lucrative oil business.

'Bumper crops, that's what I've given him. Bigger fruits and more of them. Aside from all this good fortune, I'm also celebrating my good health,' he said to me on the phone.

He had recently undergone a rather complicated operation on his right eye, which had, fortunately, been entirely successful; all anxieties about blindness or visual impairment alleviated.

'Yes, I'd heard about it and have been trying to contact you. Neither of your phones are functioning.'

'Blasted nuisance, the mobile dropped out of my shirt pocket and fell into one of the clay trenches where my vegetable beds are placed – wait till you see the size of those beauties, Carol! I didn't notice it and drenched it along with the marrows and broccoli when I was watering. I have a new phone and a different number now.'

'And your wife? Is all well at home?' I always held my breath after such an enquiry. Cécile was hitting eighty-seven. (René joked that he was her toy boy.)

'Oh, nothing wrong with her. The blasted squirrels have been a problem though. They ate right through the France Telecom cables outside our bedroom window. It was weeks before I discovered it. Just assumed no one had made contact or that Cécile's hearing was getting worse.'

After a mere five weeks' convalescence beyond his resoundingly successful optic-nerve operation, René had returned to work, masterminding Raymond's olive farm once more.

'It has grown into such a mighty concern that my presence has become indispensable.'

'Well, congratulations, René, on all fronts.'

'No, but it's getting too much for me now. I want to think about taking retirement next year when I hit eighty-five. We've employed a manager. Raymond has built the fellow and his family a house on the estate and I am teaching him everything, passing on my many years of expertise. Do bring your camera,' begged our silver-haired friend. 'My vegetables are a sight to record. I have tomatoes the size of footballs. None weighs less than three kilos.'

I agreed that we would bring a camera, though I recognised yet another of René's tall stories.

'You have to photograph the olive trees as well. Last winter, we delivered fifteen tons of olives to the mill. We are expecting an even heftier crop this year. Wait till you see the olives on them! How the branches are bowing low. It's a sight to behold.'

'Are you still spraying?' I asked.

'Of course. Why wouldn't we be? And we'll be watering all the trees old and young on a twice-weekly basis.'

This was completely unnecessary and a saddening misuse of water.

Raymond had sourced such rich flows beneath the surface of his land that he simply did not know what to do with it all. So he had dedicated a substantial plot of land to René's spectacular vegetable garden, and they had been extending, yet again, the olive groves. On top of the original five hundred older fellows, plus René's suggested two hundred trees – the returns for which had been staggering – since I had last bumped into them at the mill, another one hundred trees had been added to the groves in the early spring.

'We have our very own olive industry. Eight hundred *oliviers* now; what do you think?'

'Splendid.'

'Raymond has no interest in making money out of all this. He has more than he could ever spend. He is, after all, eighty-four years old. Precisely the same age as I am. Well, a few months younger, and he always says he has to do what he is told because I am the senior partner in this team. But I'm making a centime or two, I can tell you.'

I was straining not to burst out laughing.

'So, we'll see you next Friday, then, for a barbecue.'

*

The family living next to Quashia's cottage, she who had been so abrupt with me in the lane before the party, consisted of a curious ménage à trois: Moroccan mother, French-Moroccan teenage daughter and a French male, thirties. Within weeks of their arrival, they had set about chopping down a dozen pines and oaks on their land to make way for a marquee. Now Quashia climbed the drive with more worrying news. It seemed they had turned their attentions to a lone olive, a fairly ancient fellow that grew on a small rug of land between their fence and that of Quashia's (our) cottage. This no-man's-land was the entry to a steep track that led to the valley where the water houses for both properties were situated. Over the years we had harvested and cared for this solitary although it did not belong to us any more than it did to them.

I went in search of the new mistress of the abutting establishment anxious to return a little olive harmony to the hillside and found myself face to face with the 'boyfriend', though to which of the women I was not entirely sure. He was on his haunches outside their wooden gate in drainpipe jeans, polishing a pristine Harley-Davidson. To the right of him lay puddles of fresh creamy sawdust. I took one look at the tree. He had lopped off one of the olive's forked trunks entirely and had left the other three-quarters slain. The tree was now lopsided, deformed. Because it was an olive, I did not doubt that it would rally but the damage wrought would take several years to amend.

'I'm sorry to trouble you.'

He rose to his feet, unsteadily. His breath was like that of a wild beast. His bleached hair fell in a slick over his forehead. His eyes were set deep, blue and furious. The cut of his face, vulpine-featured, reminded me of Johnny Hallyday.

'I wonder if we could discuss the olive tree.'

He was bemused, unfocused, drunk. His response was to threaten to slay our animals if they set foot within an inch of his fence. They owned a Rottweiler, he informed me, thrusting his arm vaguely back towards the junkyard of garden where a Mike Tyson of a beast was padding about, enclosed behind a meshed fence. To and fro it trotted, encaged, spoiling for a fight.

'He has been trained to disfigure all canine relatives and competition.'

'I see. Well, I assure you I won't let our three anywhere near him. I only wanted to discuss the olive tree,' I attempted again.

'Fuck off.'

'Sorry?'

'You heard me, lady.'

I saw trouble ahead.

We had hit a spell of almost intolerable heatwave temperatures; *la canicule* is the French term for what in English are the dog-days of summer. On the date prearranged for our lunch with René, Michel and I drove through upland country lanes to Raymond's farm. His property had not been constructed out of limestone mountainside as ours had. It lay on flatter ground, gently sloping, like the spill from a higher lying plain. The placement was a good fifty metres in altitude above ours and it was possible to drive the twenty minutes' distance winding into deeper countryside, buried away from the busier routes, and find yourself in a cooler, more refreshing climate.

Raymond's property was fenced in with white pickets and resembled an American ranch. It was parkland, a greensward without terraces. This gave the place a more modern feel, in sharp contrast to the spread of our labour-intensive terraces hewn from rising acres that bore the prints of the traffic of generations upon them. Our site was of soft limestone, trodden by innumerable feet. No gentle gradations and no flat areas except where the drystone-walled terraces had chiselled them. Water tumbled southwards at our olive farm, a swift downhill flow, but Raymond did not have gravity working for him in the same way, so he was obliged to invest in expensive pumps, waterspouts, a gallimaufry of plumbing where every conceivable piece of machinery or equipment required was present on his land.

As soon as we stepped from the car, we were taken on a tour by René who was wearing a black patch over his right eye. Raymond had not yet shown himself. René said his colleague was sleeping. 'He might be my age, but he doesn't have my stamina, can't tolerate the heat or the pace.'

Raymond, tycoon turned water diviner after his retirement, had detected three generous, ceaseless *sources* beneath his fields, all of which he had deployed. Once the drilling had been completed and the wells had been opened up, they had each been furnished with an electric pump. Ground-level piping was everywhere. Several of these fed or were fed by what Raymond claimed was 'the largest private water basin in Provence'. Built by him, it held ten thousand cubic metres, and all this water was free. He paid only for the electricity. It was a complex and ingenious system that would have cost the average lay person many thousands of euros, but Raymond's father had made his fortune in agricultural machinery and hardware and even today the family owned the largest outlet in southern France, supplying farmers, industries and individuals with all that was required to irrigate, plant, construct, sunbathe, barbecue, build swimming pools – every aspect of living or working in the South of France.

I did not record the precise number of gallons that Raymond, when he did appear, proudly boasted he was drawing up on a daily basis. It was a staggering figure. Since I had returned from my travels, I had begun to appreciate a little of the water and environmental issues facing the Mediterranean and I perceived Raymond's usage as extravagant. Advised by René, he was rocketing thousands of gallons of water fused with pesticides everywhere. The residue from these settled on the ground and seeped beneath the soil back down to the flows from mountain rivulets making their way to other homes, springs, lakes, reservoirs.

'Quite an industry, don't you think?' cried Raymond.

I nodded, deeming it inappropriate to comment. Fifty-five per cent of all French groundwater has been found to be contaminated by pesticides.

'René's allotment', a quilt of land that extended for more than an acre, was bursting with exaggerated foodstuffs. The cabbages and greens were as tall as our silver-haired guru himself. He walked in among them while we photographed the sight, and he was barely visible. But his one free eye glinted with pride, with the knowledge that he was earning himself a tidy packet, selling most of this on for cash. Tax free.

'I'm building the foundations of a better life for my grandchildren. I've fallen upon a goldmine here, Carol. I only wish Raymond and I had caught up with one another again years ago,' he confided, 'but in those post-war days nobody wanted olives. The idea that they would become a luxury food item was risible back then when I set off inland to Provence *profonde* and herded goats and sheep in the mountains. It was a hard living, a pinched existence, even if I did know it inside out. Goats and olives: no one understands them better than I do. My old Catholic mother made herself a set of rosary beads out of olive stones. She prayed day and night, fingering them so hard they grew shiny as marbles. Well, her prayers for her only son have been answered, *n'est-ce pas*? Who would have ever thought my fortunes would deliver me here?'

He was by now giving us a tour of the 'bungalow' constructed for the new manager.

'What a pad, eh, what a life! I'd have brought Cécile and our boy. We could have moved in here, if only I'd known. I'd have run this place for him while he made his fortune with the agricultural equipment. He wants for nothing, does Raymond. Do you have any idea what he is worth?' René rubbed his cracked fingers and thumb together, dry as an old scouring sponge. 'Millions,' he muttered. 'Millions. But I'm the one who knows the trade. He needs me. I know olives like no one does. And Raymond's got the itch for them. He itches with it. Like you do, Carol, and that can't be bad for me.'

Later, when I complimented Raymond on the new house, he offered to put us in touch with the architect and his team of Portuguese masons. 'They are excellent, as you can see, and they are very reasonable.'

We sat drinking champagne in the shade, looking out towards a distant hillside church set upon mauve mountains while the meat grilled by Raymond, aided by Michel, sat on plates on the table getting cold. Finally, exasperated, he called into the kitchen to his housekeeper.

'Charlotte, what is going on? Where are those splendid potatoes and the stupendous broccoli all grown by our good friend here, René? Get a move on, please. We are waiting.'

A croaked voice from the kitchen returned, 'They're not cooked yet, sir, there's an airlock in the pipes and I've no water for boiling.'

To repair the leaks, to restore the subsiding garage involved shoring up its interior with pillars beneath a new ceiling. Above this, on the exterior, a damp course was required beneath a bed of cement. Michel suggested we lay terracotta tiles on the cement and create a patio. It would begin beyond the French windows of our bedroom and reach across the stables and measure approximately eighty square metres. Too expansive for our needs, at some later date, after planning permission, when we could raise the necessary funds, I imagined we might knock through the far wall in my den and build on to the tiles an extension that would supply us with one if not two extra bedrooms for Michel's daughters and grandchildren. However, first things first. The preparatory work entailed the removal of the corrugated asbestos sheets covering the stables.

We telephoned the architect who had been responsible for Raymond's staff house. A young fellow on a motorbike in his early thirties, he spent an hour with Michel, confirmed the urgent nature of the structural repairs and then informed us that, as projects went, ours was not sufficiently important for his firm. 'We usually build houses or renovate chateaux. You don't need an architect. I'll give you the names of the masons who carried out the work up at Raymond's place. They're a good team and can handle this without me.'

We called the recommended outfit, a family business that comprised a quartet of Portuguese. Coincidentally, they had tiled another portion of terrace for us, completing a job left incomplete, quite some years back while they were under contract to another *société*, fronted by one Bolmusso. Our memory of this previous experience was that they were reliable. Michel asked them to pay us a visit and take some measurements. One of the four turned up, rough at the edges, crinkle-haired and rather good looking. He took one look at the works, shook his head and said they would not touch the site until the asbestos had been removed.

We perfectly understood that, but could they not organise the removal for us?

He shook his head again. It required expert handling.

'But by whom?'

'There are companies who specialise in the removal and disposal of *amiante*, but it will not be cheap.'

He scribbled down the phone number of one such firm and said to use his name. 'Tell them José recommended you.'

I took the card and we thanked him. In the meantime, would they consider giving us a quote for the work?

'When the sheeting has gone.'

We left it like that and promised to be in touch.

As he was settling at the driving wheel of his white Kangoo, I asked, 'Don't you work with Bolmusso?'

'Once upon a time, but not any more. We're an independent outfit now.'

I telephoned the number José had given us and learned from a recorded voice that the company was closed until after the summer.

When not in my den, I whiled away the hours logging the changing seasons. I watched second-brood magpies learning to fly, their back feathers standing upwards like fans as they skittered about, took off and tumbled, dropping on to the trampoline-soft, pine-needled earth high up the hillside, rolling carelessly like dusty tennis balls. I watched water endlessly, gazing into puddles where the morning irrigation had spilled, mirroring drops, or I gazed upon the surface of the swimming pool. Like a sailor on deck, I studied the sky's reflections, a lone cirrus cloud passing, or the swirl of movement when a fly, a bluebottle, a bee landed and circled for its life until I waded in to save the unfortunate insect. Then I lay dripping in the garden, having abandoned my weeding or reading, and stared at the sky itself, considering its ineluctability, its lack of seams, wondering at how it never got torn, its ability to span space, to knit so perfectly with the horizon. I talked to lizards, too fearful of me to converse. Still, I pursued them, peered with a giant's perspective into the schisms, hairline fissures within the rocks or tiled surfaces, the cool, dark shadowy places beneath terracotta pots that had become their homes. Umbras I could never penetrate.

My mind was forever busy, but it was also a hollow, troubled space. The olives were growing fat, though still hard and green as peas. Quashia was chuffed by the quantity weighing down the trees and constantly reminded me that, had he and Michel followed my guidance, there would be few that had not been tainted by flies. I did not want to leave here but the poisoning taking place on the hillside had become untenable for me and I had to do something about it. Either find us another home without trees requiring chemical control or get busy and dig up alternatives. Somewhere, there had to be a natural enemy of our fly, but where? If no one else had succeeded in finding a solution, how was I to?

France ranks third in the world for its agricultural exports and is number one in Europe. Approximately one half of its entire landmass, which is to say twenty-eight million hectares, is dedicated to agricultural production. This had been achieved by three and more decades of intensive farming and a far greater use of chemicals. Chemical-control products have become easier to obtain than ever before. Today, France ranks third in the world for its pesticide consumption and is the leading user in Europe. But the France I had returned to was in change. It was considering its ecological responsibilities, reassessing the negative impacts of such farming.

My adopted home was under the leadership of a new head of state, M. Nicolas Sarkozy. On the evening after his election, he launched *Le Grenelle de l'Environnement*, an Environmental Round Table, an open forum for debate, consultation and action plans for the future of agriculture and its effects on the environment. The themes under discussion numbered six: climate change, biodiversity and natural resources, environment and health, sustainable production and consumption, ecologically responsible democracy, green development employment and competitiveness.

I was not a fan of Sarko, but, when I had learned about this from Michel during my travels, I hoped it would bring the issues out into the open, create dialogue, admission of responsibilities and enable shifts towards positive changes and a greener, safer world.

September now, beyond my return. The second of our ancient cherry trees, the widow of a pair, was showing signs of ill health. Its leaves were turning their habitual golden apricot.

Against its chocolate-dark bark, it was a glorious sight, but there was so little foliage. The tree had delivered no cherries this last spring, Quashia and Michel confirmed. I hated the prospect of losing it as we had its partner so we lopped off its sparsely furnished canopy and left it, with a trio of thick, naked branches rising out of a mammoth trunk, hoping that, when spring came round again, it would reshoot.

The men were notching up our third spray of the season, and the fruits were looking rotund. I turned my attentions to the challenges of asbestos removal. Now that August was behind us, *la rentrée* was beckoning and the hordes of tourists were driving back up north. The roads were clogged in every direction and businesses were back at work. Schools were opening their doors and I was at last able to get responses to my calls. Someone from the asbestos-disposal unit of the *société* recommended by José finally picked up the phone. Yes, the woman confirmed, they specialised in asbestos removal.

I requested that they drop by and give us a quote.

The section of cover in question measured five metres by three and a half. In the old days, before we lived here, these *boxes*, stables, housed four horses, I had learned from Quashia. This was an insignificant day's work, or so I estimated, but the first quote we received came in at a whisker under seven thousand euros.

Seven thousand euros!

Michel shook his head in disbelief. 'They have made a mistake,' he said. 'Or they are crooks.'

I called the gentleman who had visited us and queried the figure.

'Surely you have muddled us with another job?' I said.

'No, the treating of asbestos is a costly business. It needs to be removed, laid out on a wooden palette and wrapped in protective sheeting, the thickness of which is crucial. Then it is taken away by lorry, registered with the toxic material office, deconstructed, disposed of and you are given a certificate of clearance that must be filed with your property papers at the notary's office. The imperative in all of this is the certificate. Without it you are forever responsible for those few sheets of asbestos.'

'Why does it have to be so complicated?' I sighed.

'It's a toxic material. However, off the record I can probably find you someone who'll do it cheaper. A cash arrangement.'

'Do we still receive the required certificate?'

'Of course.'

'Then please, Monsieur, send us whoever you recommend around and ask him for a cheaper quote.'

'It'll be done tomorrow morning, Madame. Count on me.'

I never heard back. I telephoned this gentleman twice more and each time he made the same promise. I scoured the Yellow Pages and found one company advertising such a trade and I phoned a number I saw advertised on the side of a van parked in the street. Both acknowledged that they were in the business of asbestos treatment but they would not work with private individuals. The contract had to be drawn up through a registered company.

I telephoned the local town hall and was connected to the planning office who furnished me with two numbers, warning me that there were only three companies still in existence which offered this service. The third was the one I had already had dealings with. The other two I contacted. The first no longer handled dangerous materials but recommended the firm which had given us the quote for seven thousand euros and the second passed me on to someone else who promised to call me back and who, in spite of several reminders, never did.

We were back to square one and we had little confidence in the garage roof seeing us through the winter.

Michel was more upbeat, more determined. He suggested we track down Bolmusso, the original boss of the Portuguese artisans. That same evening, Bolmusso crept stealthily up the drive in a spanking new, state-of-the-art four-wheel-drive. He was older than I remembered, of course, and rounder, but, yes, I recollected him now; less hotheaded than his erstwhile Latin employees and craftier. He had tricked me over a delivery of bathroom tiles, charged us twice. Still, Bolmusso, heaving as though suffering from emphysema, prepared us a quotation and added on to it every conceivable and inconceivable extra. To repair the garage roof, the *devis*, the estimate, he sent through was a snip under seventy thousand euros.

Seventy thousand euros!!

And the corrugated sheets?

He shrugged. He could not help us with asbestos.

We requested bids from four construction firms to give us a more global picture. Bolmusso's was the highest by more than twelve thousand. He was, though, the only one of the companies that had been employed by us before and, bathroom tiles or not, we had been pleased with the masonry work. However, as Michel pointed out, it had been the Portuguese family who had actually executed the work and who had also delivered such excellent craftsmanship *chez* Raymond. Michel telephoned Bolmusso and requested that he '*fait un effort*', make an effort, to drop the price, which he did. The result was sixty-eight thousand euros.

It was out of the question and we left him a message telling him, *non*. Regretfully so.

October. Was it possible that the government and I were in step, I joked to Michel over breakfast. An ambitious action plan, seeded by the results of Sarkozy's Round Table, had been presented by the French minister of agriculture, M. Michel Barnier. The goal was to reduce the use of pesticides by 50 per cent before the year 2018, increase the surfaces dedicated to organic farming by up to 20 per cent before 2020 while commercial production of pesticide-resistant GMOs, genetically modified organisms, specifically food crops, was to be suspended altogether pending new investigations beyond the creation of an investigative body. This was a major coup for all those who had demonstrated (numbers had gone to prison for their convictions) against genetically modified crop plantations here in France.

'This is very optimistic,' I said to Michel.

The investigative enquiries were to be achieved by the setting up of research programmes and motivating institutions dedicated to seeking out more ecologically responsible alternatives, relying less on the pesticide industry while encouraging farmers to change their approaches, to learn, to understand the sweep of their personal responsibilities.

'It sounds optimistic,' agreed Michel. 'However, it is all open to further lobbying. Nothing has actually been implemented yet.'

I, meanwhile, had read about a variety of chrysanthemum that was a natural insect repellant. Might it work on the olive fly? I had yet to find out. I could not track down its name. I began to scour ancient pest-control methods, classical texts, Latin for the most part, agricultural tomes written during the heyday of the Roman Empire. Page after page I printed off the internet. My Latin was hopeless. I had left behind even its basic vocabulary more years ago than I cared to admit. Trying to make sense of it was frustrating. Then, into this haze of concentration came a call.

'Madame?' a voice rasped like the low, slow slice of a chainsaw. 'Let us come and see you. *We* will make an effort.'

'And you are?'

'José.'

'José? From the Portuguese team?'

'*Si, Madame.*'

This did not sound like José.

'Tell your husband that I and my brothers – it's a family

business – will come and talk to him.' His was the voice of a gangster from a Hollywood Mafia film.

However, they arrived. Four of them, José, José, José and Francisco, the family team of Portuguese masons. Francisco was the brother of one of the Josés, though of which we were never quite sure. The other two, José and José, were associates, perhaps cousins.

'There are no bosses among us. We operate as a team. All four of us are masons *par excellence*. As you know, Madame, Monsieur. We'll better whatever quotes you have received.'

'And the asbestos?'

'We'll find you someone to remove it.'

'A legitimate certificate is an imperative.'

'We'll make sure you are provided with whatever is required.'

They drank wine, shook our hands and left. We never heard back from them in spite of several reminders on my part.

'José?'

'No, Madame, this is Francisco.'

'What about our asbestos?'

'José is dealing with that. Call him.'

I'd dial one of the other six numbers on their card. 'José?'

'Yes, Madame?'

'The removal of the asbestos, what is happening?'

'José is dealing with that.'

'But which José?'

'Big José.' I could not for the life of me remember which was the 'big' one nor which of the phone numbers would put me through to him. In any case, I never reached him. If I did, he must have passed me on to one of the others. The fact of the matter was that winter was approaching and we were getting nowhere.

If we could not legally remove these sheets of asbestos we would be unable to extend the terrace and that would mean no groundwork for the bedrooms I had been dreaming of, no child-friendly facilities for Michel's family. We were snarled up between the toxic materials departments and masons who saw this as a money-making opportunity and no one could point us in a direction that led anywhere. Should we shell out the seven thousand euros and get matters moving?

In those weeks preceding the harvest, when the Latin research grew too painstaking and while we awaited feedback from the Portuguese, I haunted the garage, sponging up its ever-worsening leaks, mopping away the occasional puddles, slinging out the broken tiles, plaster dropping from a ceiling that had deteriorated to the point of baldness, spent hours among its increasingly fragrant mildews, where gardening tools, land equipment and spraying machine were stored. I packed and repacked material in attempts to protect everything during the upcoming wet season.

There I came across the residue in a ten-litre plastic container of the insecticide we had been pumping into the silvery branches throughout the summer months. I lifted this substantial *bidon* out of its cool-storage home into the late autumn sunlight and took a long, hard look at the neutrally toned liquid and its directions for use.

The product that we were annually spraying on to our trees as recommended by the olive bodies, including the Chambre d'Agriculture, stated on its label that, after a period of four to six weeks, no residues remained in the soil. This implied that none of its poisonous properties would sink beneath the earth's surface and find their way into the groundwater. Groundwater, underground mountain springs in our case, that feed local reservoirs and are eventually piped into the taps of every modern home. These were not at risk, the label claimed.

I wondered.

Its discharge was safe for agricultural purposes, it further stated, and it was compatible with numerous other insecticides and fungicides. The product's makers, however, *declined to accept responsibility when it was used in conjunction with other products unless advice had been taken.* An interesting caveat, I thought.

Towards the lower section of the label, in small black lettering, I took a magnifying glass and read the following:

Produit dangereux pour les abeilles et autres insectes pollinasateurs. Ne pas traiter pendant la floraison.

Translated it warned, 'this product is dangerous for bees and other pollinators' (which would include butterflies as well as certain birds, wasps). 'Do not use during the flowering season.'

But the flowering season of which plants? There are flowers of one variety or another in blossom throughout the entire year. Olive flowers, appearing in April, are never sprayed. Their arrival precedes the drupe and the fly that lays its eggs in the drupe's flesh. The drupe or olive appears a month or more after the flowers have died away.

So, which flowers were the manufacturers referring to?

Further research on my part brought forth other results that were even more perturbing. Dimethoate, the principal ingredient in this spray, disrupted the endocrine and nervous systems in both humans and animals. It caused infertility. It was sold as an insecticide to combat, to kill off fruit flies that inhabited asparagus, cherry, chicory etc. and olives, and its duration of efficacy was two to three weeks. At which point it was necessary to spray again. I considered my own childless state and shoved that thought immediately out of mind. I wondered about the nervous system of the honeybee. François had told me so long ago that bees were being wiped out because the network of cells within their systems that communicated navigational information was being damaged by contact with insecticides. This caused them to lose their ability to find their way back to their hives and, in their tireless attempts to make it to their dwelling, they wore themselves out and died.

Residue of this product, dimethoate, on the fruits it had been used to 'protect' could not be simply washed away. This explained the reason why, of course, no harvesting was allowed until at least twenty-one days beyond the final coating of the season. Was the company responsible for this product, and possibly other products, unaware of the longer-term effects of their fly poison, or were they disregarding their responsibility towards the ecological systems that were being devastatingly disturbed by the use of concoctions such as theirs? I made a note of their name and address.

The official argument was that twenty-one days after the liquid had been used it was no longer efficient. Consequently, twenty-one days after the ultimate spraying of the season all risk had been eliminated and the olives could be harvested without delivering any harmful effects. Of course, any pollinator waywardly frequenting an olive grove, happening by, busying herself feeding off rogue flowers during the time of spraying and beyond, had long since been done for. The product did not exclusively target the olive fly. Its efficiency was indiscriminate, non-selective.

Two of the Portuguese returned, Francisco with his brother, crinkle-haired José. They met up with Michel, reconsidered the site, apologised for not having found a solution for the illegal roof sheets but agreed to be 'more reasonable' with their estimate than their competitor Bolmusso had been.

Their quote was a little over fifty thousand euros. This remained beyond our range. However, after Michel telephoned and told them 'No, thank you' they began ringing us twice a day.

'No, no, you haven't understood, Monsieur, that is only an estimate. It is the price before the reduction. It is not the real price. We can discuss the real price.'

I think we had forgotten this slice of Côte d'Azur life.

*

Michel had long been in the habit of acting as a newscutting service to me because he felt I did not spend sufficient time reading the daily papers (which was true). Today's I found on my desk, an article from *Le Monde*. It reported that the French agricultural minister, Michel Barnier, had informed his parliament that bee deaths were being confirmed in areas of France where the insecticide product Gaucho (and another known as Regent) had not been sprayed. The intimation being that neither Gaucho nor Regent could be held responsible for the current mass disappearance of French bee colonies.

The beekeeping associations were losing their case. I tossed aside the article with a sense of anger and impotence.

Were these reported deaths in areas of France where fruit orchards were farmed, where the crops were managed with chemicals? What could be done to help the bees? Of course, it was too late to assist François, our dear friend who had invested all his and his wife's savings in his dream of hives and honey. They had found themselves victims of what was being described as a 'mysterious international bee crisis'. But was it a mystery, or might that possibly be a high-level smokescreen? And what of the prospects for our bankrupted friends in their retirement and for the future of the honeybee itself? A world without bees? Personally, I missed the company of all those furry little foragers flitting busily about our grounds and I was beginning to wake up to the fact that there were far fewer butterflies, songbirds, insects and pollinators frequenting the farm. My memory was that, during our first years, I was constantly surprised and delighted by sightings of yet another new creature, but they were less in evidence now.

I drank my morning coffee, deliberating, and decided to contact the producers of dimethoate, whose registered offices were in Belgium. I looked them up on the internet and accessed their site, but to download or read any information involved the disclosure of a certain amount of personal information on my part: identity, purpose of my business, reason for contact, etc. This I offered and followed it by the subject of my enquiry: bees and the environment. An interesting caveat before SEND informed me that I had a right to protect or alter my personal information or object to it being forwarded to third parties, but in order to do that I was obliged to send in a written application to a given address in Paris. I decided not to bother and pressed SEND.

We began gathering our olives in the second week of November. The fruits had ripened ahead of schedule, as had happened during the previous couple of seasons. It was becoming the norm, but I had not noticed any real expressions of concern within the agricultural community. I drove to the Var, stopping randomly to chat with olive growers, pickers, crofters, smallholders. I wanted to hear the millers' opinions, too, but the majority were still closed. I was keen to sound out the general reaction to this gentle season displacement, to get an idea of how many mills were opening early to accommodate Nature's cycle transition. Few of those I spoke to, folk out on their lands, appeared to be unduly worried, few mills were as yet turning. I found one ready to go. The prelude to the season of olive gathering has always historically been linked to the last week of November and most were sticking to that.

'Why change?' was the response I received. 'This is abnormal. Things will be back to their regular ripening period next year.'

'But our fruits,' I argued to this farmer or that, 'save for any lost crop, are encountering a third successive year of premature ripening.'

'You know why, don't you?'

I shook my head.

'You probably haven't sprayed them hard enough. It's essential to soak them in insecticides. That keeps them healthy.'

'Does it?'

Ours was a reasonable crop that autumn. On that score we had nothing to complain about and Quashia was happy that I had not denied him a harvest. We picked and gathered fast with nimble, experienced fingers to bring in a fresh yield, low in acidity. As was the tradition *chez nous*, I was the one to transport the fruits to the mill where the oil arrived, spluttering and gurgling like a newborn child, and its quality was good.

Early December wiped her muddied feet on the mat and I was still shunting olives to the mill.

'How are you doing?' 'How was your year?' Standard questions for those who meet but once annually at the pressing station.

During my penultimate appointment, a certain festive mood had taken over, warming the cold subterranean air where the exhalations of breath rose like smoke clouds and men cradled their wine glasses with gloved hands.

'Not too bad. Can't complain' (which most usually did).

'How was it at your place?'

'Here, I brought a bottle of my own wine. Will you join me in a glass?'

'Won't say no, *merci beaucoup*.'

I was by now pressing our third load of the year, while listening to these exchanges – three days' pickings, fourteen crateloads, most of which had come from the lower groves, from the revered knobblies, givers of plump, oleaginous fruits – when I bumped into René and Raymond. Here they were together on that morning celebrating their weekly haul of a minimum of one ton of olives.

Apropos of not very much (besides long-lost friendships and the enduring power of the olive), I was invited to share a bottle of rather splendid rosé with them. Even though it was only nine in the morning, we settled contentedly together to a glass or two, as is the habit on a wintry occasion while waiting for one's fruits to be metamorphosed into luscious golden oil, and we fell into conversation about the state of the planet and, quite naturally, olive farming in particular. I knew well that René and Raymond were hardline conventionalists. Their thoughts were clearly set on their annual haul. For the world beyond their fences, they could not have given a hoot. Quantity was their boast and it almost matched their, it has to be said, exaggerated claims for the quality of the oil. Already this winter they had harvested five tons, or five thousand kilos of olives, and still had approximately twice the same quantity to go. We at Appassionata own three hundred trees, a dominance of which are juniors, and we usually shunt to the mill a total of around twelve hundred kilos of our purple-green drupes. Raymond, with all his recent plantings, owned close to three times our figure, including his juniors, yet his yields were thirteen or possibly fourteen times greater than ours. Their boast was that they had irrigated the fields on an almost daily basis. Old trees, young trees, it made no difference; each plant was being fed with phenomenal quantities of water pumped from Monsieur le Water Sorcerer's trio of home-sunk wells. And on this particular morning I dared to pose a question that I knew would be a stick of dynamite within this crowd of farmers whose faces were the colour of baked bricks.

'Raymond, even given that the water beneath your land is free, are you not just a little concerned about the overuse of natural H_2O resources? After all, olive trees don't require such a litreage of irrigation . . .'

'Heavier watering produces heartier harvests.' A recurring mantra of René's these days, since he had gained liberal access to Raymond's gushing wells. He poured some more wine, refilling all our glasses, but I could see he was not happy with where I was leading the conversation.

'I was in Andalucía a little while ago and, while I was there, I had a conversation with a soil scientist.' I did not allow the frowns gathering like bad-weather clouds across the foreheads of those present to intimidate me. 'He was an intelligent, caring being, born of the earth with a grand lineage in olive farming and he expressed profound concern for the damage and erosions being wrought upon his southern Spanish soil. He talked of desertification—'

'Desertification! What in God's name is that?' spluttered one bloke with sagging, turkey neck and a barrel chest.

'It's the reduction of fine farming earth to desert-like conditions, when the soil becomes arid and lacks all nutrients, including water.'

'Let the Spanish worry about their own problems,' rejoined René. 'There's no likelihood of desert conditions here!'

'Ha! Ha! A desert in Provence!'

'Certainly not!'

'What a load of codswallop!'

Hearing the outburst, two other farmers, younger men, shuffled closer. Another generation: perhaps a more modern approach? Both were dandling glasses of a local heavy-bodied Côte de Provence red, poured for them by Gérard, our mill owner, who was across the cold room bending and shunting, working non-stop to assure our loads were kept on the move, down the conveyor belt and into the crushing baskets.

I yearned to share further concerns, to elucidate upon ample and rich discoveries of mine with these newcomers to the debate as well as with my old pals, René and Raymond, but I sipped my wine silently, deliberating. For well over a decade René and I had been in disagreement over our approach to olive farming and I itched to attempt just one more time to persuade him to think again, but he was recounting to the pair of farmers who had just joined our little circle the gist of our conversation. When he had finished, the two men looked my way, raised their eyes in contempt, muttered words such as 'she's nothing but a foreigner' and, from the other, 'and a woman', and they turned their attention to

Raymond, who was rather out of his depth in the bowels of the mill in his tailored slacks, cashmere scarf, sweater and hunting jacket. The farmers would undoubtedly have known of him – 'a wealthy man, that one. Inherited his fortunes from his father's business acumen, of course –'. Such whispered comments might have passed between them elsewhere but here they both stared and nodded deferentially. In any case, Raymond was new to the world of the olive and simply repeated whatever information he had been fed by René. These recent triumphs lay in his farm's phenomenal production and Raymond was basking in this new lease of life, of glory. I decided to keep any other opinions and travel experiences to myself. Instead, I raised my glass to wish them both, him and René, good health. '*Santé!* Here's to a continually robust harvest.'

Debating further was futile. Why spoil a decent friendship when I knew they were not to be swayed?

Ironically, though, it was through René, by chance on that Saturday, that I learned of the existence of a certain vegetable farmer, a local resident and maverick who went by the name of Luke. He lived in the hills behind Nice and was 'trying to change the world'.

'Just like you, Carol,' my silver-haired friend quipped. 'More concerned with saving the planet than producing quality olive oil.'

'In what way?' I asked.

'Thinks organic,' pronounced René, spitting out the words as though they were a disgusting taste. 'A couple of pals were talking about him, read about him in the farming gazette, some darned trick he's got up his sleeve.' Luke – René gave me his full name – intended to outsmart the chemical companies.

By now it was late morning and a full crowd had encircled us, listening in, greedy to hoover up other folks' business, for there is not a great deal to do while the paste churns, a process that takes a couple of hours, and is transformed into oil. They began chipping in, these twisted and ruddied faces of the earth with hands as rough as bears' claws, haranguing me with their heated opinions, nosing into our conversation as was their norm. One farmer's business down here was every other's.

'And why not, why shouldn't he try to be organic?' I protested.

'Because he won't succeed.'

'You can't,' was another's opinion.

Bah and humbug was the prevailing attitude of those standing about me.

'This woman here is under the impression that Spain will soon be transfigured into a desert.'

'For heaven's sake!' cried some.

'All the better!' cried others.

'You've got to kill those little blighters off good and proper, poison the creepy-crawlies or they multiply. This bloke – what's his name? Luke? He's not doing any of us any favours. The infestations spread and then we all cop it,' berated another, spittle dribbling until trapped within the curvature of his upturned, stubbled chin.

'There are no solutions that work, that are *efficace*, effective.'

'Only pesticides do the trick.'

Mmm. I sipped my delicious rosé, enjoying the illicit experience of being one notch tipsy at ten in the morning, and kept all further opinions to myself. The subject was a minefield, but I had made a mental note of Luke's name and how I might get hold of him.

Tracking Luke on that first occasion was perfectly straightforward. His telephone number was displayed on the half-page of advertisements in our weekly agricultural paper. I rang it, left a message and he returned my call within a couple of days.

Michel suggested we invite him to lunch and find out what

his 'maverick' ideas were all about, which we did and he accepted.

Luke became my hope, my beacon on a precarious landscape. He was a *petit fermier*, he had told me on the phone, predominantly a *maraîcher*, a cultivator of vegetables on a modest holding that also counted a few peach, apricot and almond trees. Olives were a small return for him, little more than a sideline. However, he had taken a step, which, when we first heard about it, I had judged to be bold, defiant and ahead of its time. He, like me, had grown tired of waiting for an alternative to pesticides to become available and had taken matters into his own hands. To defeat *Bactrocera (Dacus) oleae*, Luke had called in the help of a smart little foreigner.

If this was to be believed, he had achieved the impossible. However, fly populations were greater in the hot, humid coastal regions such as ours so Luke would have had a slightly easier time of it at his place.

Bactrocera (Dacus) oleae (nicknamed 'Dacus' on our farm because it was less of a mouthful) was a small fly about five millimetres in length. He hibernated in winter, buried within the soil in Mediterranean olive groves, and only surfaced for procreation purposes when the weather grew warm. Twenty-five degrees Celsius and upwards was his happy zone. Dacus' female partner, when she was ready to give birth, infested the silvery foliage and inserted her ovipositor, her egg-laying organ, into budding olive fruits, piercing the skin's green surface and leaving a small black circle; her mark of entry, her passport stamp. Within two to four days, her eggs had hatched into larvae, which began to feed off the inner flesh of the olive, gorging ravenously, sapping it of all nutrients.

This tiny black-tip-winged insect had, according to scientists, been in residence around the Mediterranean for two thousand years and was the principal enemy of Old World olive farmers. Up until now it had been unstoppable unless treated with toxins, such as the product we had sitting in our garage. Scientists, farmers, soil experts, major chemical companies, greenies – everyone who was interested in finding a way to improve the quality of olive produce, not to say the quality of life, had a vested interest in controlling this little critter, but this fly had the olive world at its mercy.

However, Luke had beaten the experts to the table. When I spoke to him, talking very fast he briefly explained the reason for his success. He had found a fly that fed upon the olive fly.

'It's that simple,' he said. 'Nature's own predator.'

Luke had released on to his land an enemy, a predator of Dacus. *Psyttalia lounsburyi* was an African exotic, a fly that had originated in Kenya and South Africa. This alternative fellow was a parasitoid; it fed off the larvae of the olive fly thus destroying all future generations of Dacus and potentially wiping our enemy out. It sounded ingenious. I was very excited.

Mid-December. I was ascending the driveway, returning from my last mill visit of the year, and spotted a man strolling with Michel through the olive groves up behind the villa. This had to be Luke. Today was the date for our designated lunch. I could not hear their conversation as I parked up and pulled out our containers of freshly pressed oil, but I assumed they were estimating the number of flies we would be requiring and an approximate costing for them. Dressed in denim, he exuded an air of casual confidence. Both Michel and I took to him immediately and we listened eagerly as he told us over lunch that he was considered a bit of an outlaw in the area, 'a renegade,' he laughed, because he had followed an EU directive that had caused one of his colleagues, a fellow vegetable farmer in northern France, to be thrown into prison.

'Tax evasion or selling rotten fruits?' I laughed.

The man's crime was that he had been caught by the police driving his truck on a public road. He was in possession of a clean driver's licence and full insurance, but his motor had been fuelled by an illegal substance.

'What might that be?' I insisted. Michel always warned that my curiosity knew no bounds.

'Pure vegetable oil,' grinned Luke.

We learned that it is both legal and a common practice to drive tractors and trucks, any diesel-powered vehicle, in fact, fuelled by vegetable oils on private land, but not on public roads, which was where his comrade had been 'nabbed'.

'I do it all the time,' continued our guest, 'but the difference is that I have not yet been hauled over. In fact, there's rapeseed oil powering my old gal right this minute. That's what got me here.'

He must have caught our surprised expressions because he grinned and said, 'Don't worry, you cannot be held responsible while my bus is parked on your gravel. This is a private estate and once I leave here, well, that's my own affair. My colleagues and I see it as both a gesture for the environment and an up yours to the state.'

Should that have been a warning to us? I don't know. I like to believe that I still possess a streak of my younger revolutionary spirit and, when the conversation eventually turned to the purpose of his visit, I was engrossed.

Luke was vanguarding another *petit* revolution down our way and I was keen for us to jump aboard. Indeed, it was not so *petit* at all.

He had been trading in vegetables for over twenty-five years – he was surprisingly older than he looked – and had gone *bio*, converted his farm to an organic one, six years earlier. His main crops were salad vegetables: tomatoes, lettuces, onions as well as aubergines, potatoes, courgettes. Alongside all these he kept a couple of goats, made cheese for his family's consumption, and owned forty olive trees up behind Nice. They, like ours, were of the *cailletier* variety.

Although he was a little further inland than us, Luke's olives had regularly been blighted and, also like us, he had grown tired of not being able to find a gentler alternative to pesticides. Until, somewhere along the line, he had come across 'this exotic species of fly'.

'*Psyttalia lounsburyi*. She's a hot, sweet, tropical beauty and a natural enemy,' he winked. He might have been talking about a mistress, a dangerous liaison.

A dangerous liaison indeed!

After learning of Luke's existence from René, before this first lunch with him I had tried to find out about this 'tropical beauty' who fed off our enemy. The fly was not some tale. It really did exist and it had been imported by scientists into France and California. *Psyttalia lounsburyi* was a natural and recognised predator of the olive fly and was, even as we spoke, being studied in a research laboratory not twenty miles inland from our farm.

'But it is not as yet out in the fields; it is not available to farmers.'

'Oh, but I had understood that you had already . . .'

He nodded. Luke had, during the previous season, released batches of these flies into his groves and had been rewarded with a splendid harvest of fly-free fruits.

'This organic alternative works,' he smiled, forking one of our greenhouse tomatoes into his mouth.

'Why are the laboratories holding back then?' I wanted to know.

It was Michel who offered an explanation. 'It could be irresponsible,' he said, 'to let loose any form of foreign life, be it plant or creature, into a territory that is unknown to them. Their behaviour patterns need to be thoroughly studied and understood. Consider the devastation caused by the introduction of rabbits to Australia and then, later again, the myxomatosis virus used to kill them off.'

I wondered where Luke had come by the flies, if they had

not yet been released on to the market, and I asked him directly.

'Out of Africa,' he winked. And if we were genuinely interested, he would import the required batches for us.

'How do we let them loose on the land?' I puzzled.

Our guest assured us that he would take responsibility for that. The insects would be delivered in long rolled nets. These 'baskets' would then be hung from a selection of trees, not all trees but one in every twenty, Luke was calculating.

'And then?'

The predatory insects were left alone. They were in open-ended containers, at liberty to come out at their leisure.

'And then what?'

'They will feed off the larvae growing inside the olives.'

'So this happens during the summer months?'

'Precisely.'

'And then what?'

'That's it. Without youngsters emerging from the larvae, the olive fly cannot propagate. Within one season, maximum two, you will have no more fly problems.'

The way Luke described it, it all seemed remarkably simple and I was ready to take the gamble. Still, in spite of my determination to go organic, Michel stayed my hand. He remained reticent. 'If the flies have not yet received the all-clear from the Ministry of Agriculture, then perhaps we should await the laboratory results,' he reasoned. 'They might discover that this exotic does not settle well here or predates on, targets, other species. Risk-assessment is essential.'

Luke shook his head firmly. 'There's no risk, Michel. This predator is safe. The change involves red tape, a complete shift in attitudes and that is what is holding everyone back. Trust me, I have tried the fly myself and it works. On top of which, I can assure you, it causes no auxiliary damage. *Psyttalia lounsburyi* feeds off olive fly larvae and destroys it and its growth potential. That is it.'

I glanced towards Michel, wondering whether or not this had closed the deal. Luke, sensing Michel's ongoing uncertainty, continued:

'Everyone is dragging their heels because going organic means that the pest-control companies are going to lose fucking fortunes. They're in cahoots with the governments, got them in strangleholds.'

It was certainly a fact that agrochemicals were exceedingly big business (approximately twenty-five billion euros a year and rising). France was the European Union's leading agricultural producer and the number one user of pesticides. Western Europe represented 20 per cent of the world's market.

'Listen, guys, go organic! I'll take care of both the shipping arrangements and *la douane*, the customs. I'll organise the transport of the little cutesters from South Africa and they will be delivered right here to your door, and I will assist you with their release. I cannot offer more.'

'Is it illegal to bring these flies into the country? Will we need a special permit?' I wanted to know.

'Look at it this way – it's a bit like the gas. The European world is shouting loudly for everyone to support a cleaner future, cut out the fossil fuels, get green, but as soon as you take the initiative they nab you for some infringement or another. If I want to drive my old truck powered by rapeseed oil, whose bloody business is it but mine? But if I'm stopped, I could well be fined. I might even get a short prison sentence.'

'Prison does not seem likely,' argued Michel. 'This is France, after all. We are not living in a police state or a dictatorship.' Having said that, we all concurred that the infringement of certain personal freedoms, the heavy emphasis on national security, the high police presence on the streets, made daily life under Sarkozy a less tenable option; the new face of France was growing to resemble a police state a little more each day.

Lunch over, we shook hands with Luke, thanked him for his time and agreed to consider seriously his proposal. In the meantime, Michel requested that he contact us whenever he could with costings and the availability of a *lounsburyi* shipment. When the facts had been laid out on the table, then we would make a decision and let him know.

I wonder now how neither of us suspected anything. I was so keen to find a new way forward that I can only suppose I was blind to the realities.

3

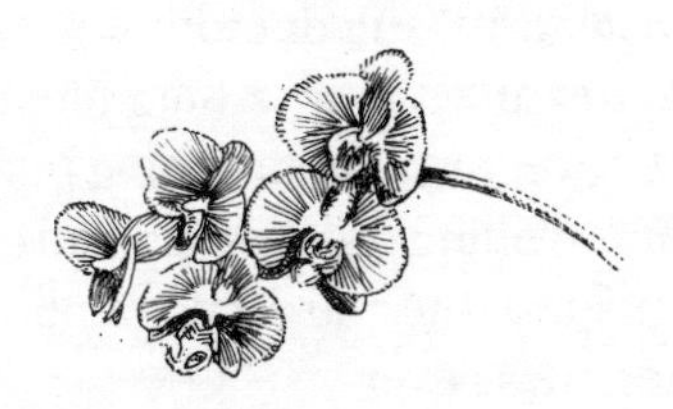

The day after our visit from groundbreaking Luke, Michel flew up to Paris for meetings and to begin preparations for our family Christmas there, while I, still at the farm, took Cleo to the veterinary clinic. Her heart since birth had always been a little faster paced than average, but on this winter afternoon it appeared to be beating at a rate that was a little alarming.

'We will need to call in a specialist from Nice,' I was informed by our vet. The robust German who had looked after our animals for years, a kindly, sensitive individual, had retired at some stage during my travels and this young lady, along with another, had purchased his business. She or her partner had cared for Bassett during his final, rather torturous hours. I had not been here, so I had little experience of these women.

'Shall I bring her back tomorrow?'

'It cannot wait. Why don't you take her for a short walk and we'll call your mobile when the doctor arrives?'

I nodded.

'Her heart is enlarged' was the specialist's verdict, three-quarters of an hour later.

I watched on helplessly, electrode conductors clipped to my beauty's belly, without grasping the gravity of what was taking place.

'Do we know what has caused this and how soon it will be before she's back to normal?' I asked.

'Ohhh, *mon Dieu!*' cried the doctor.

Cleopatra had begun to emit a high-pitched keening that chilled my blood. Save for the quickening pulse, she had been fine while we were strolling together up and down the narrow winding alleys behind the surgery, gambolling about my ankles, as affectionate as ever.

'Is she in pain?' I begged, while the two vets and one female assistant began running back and forth, grabbing at instruments and cuts of cloth. From the intense expressions on their faces, it was evident that something serious, untoward, was taking place, and then, without further warning, Cleo, our soft-furred, truly beautiful Alsatian just keeled over, collapsing on to her right side on the vet's table, the two electrode conductors still clipped to her.

'Can we unplug those things, can we, please?' I begged.

There was panic now, pandemonium. A small wiry terrier, charcoal-grey, was whining in sympathy with our girl although as far as I was aware he had never encountered her before this evening. Cleo was screaming. I was not sure that I had ever heard a dog actually scream before. Horses on the farm in Ireland when I was a child, yes, but never a dog. Such thoughts were passing through my mind while I watched on helplessly, constantly jostled from one spot to the next.

'*S'il vous plaît?*' I shouted for the third time. 'Unclip those things from her, please!' But still neither of the two women, our vet and the young heart specialist from Nice, was paying me any attention.

And then Cleo fell silent. A silence that was ominous and more vocal than her screams.

'She hasn't died, has she?' I mumbled incredulously. The heart specialist began to pound the dog as though she were making bread, kneading furiously into her fur.

'Stop it, please. Is she dead?'

My vet came and stood at my side. 'Yes, I am afraid that she is. Her heart was weak. This was inevitable.'

'But she was fine this morning!'

'No, she wasn't. You were not aware of the problem, that's the difference.'

'Well, then, leave her, please. Just let her rest.' I was fighting emotions: anger, pain, incomprehension, shock. Cleo was not even three. She had been born on the farm along with twelve others on a riotously jolly, house-full-of-guests New Year's Eve. All the puppies were pure-bred German Alsatians. Of Lola's thirteen, three had died within the first twenty-four hours. The mother herself had disposed of those tiny corpses, we had no idea where or how, and we had reared the rest. Quashia had constructed a spacious kennel to shelter them against the winter within the stone ruin above the semi-defunct vineyard over on the Second Plot. Ten puppies scooting all over the place, dribbling milk, peeing on the terraces, snouts fossicking in the earth and ripping up the flowerbeds. It had been a chaotic, harum-scarum three months but also a time of bulleting life, of curiosity and endless mischief. I had loved it, loved them all and had wanted, impossible of course, to keep the whole darned pack. Every single wagging-tail ball of trouble that they were, eight males and a pair of undersized bitches. We already had three adult dogs at that time. During those months, I was at the pet shop and supermarket every couple of days buying trolleyloads of milk, biscuits and meat. Shoppers stopped in their tracks, incredulously surveying my purchases. Eight I had reluctantly found homes for and we had kept two: Cleo and her brother, silky, black-haired Homer with one floppy ear.

And now this. Not seven days before Christmas. A couple of weeks before her third birthday.

'Madame Drinkwater, are you listening?'

'Sorry?'

'I was trying to tell you that we can deal with her, dispose of the body,' the vet was repeating to me, talking slowly as though to one who did not speak the language, while her colleague began in businesslike fashion to unclip the electrical apparatus attached to the corpse and pack her scanning equipment away.

'Absolutely not. I will just need some help, please, to carry her to my van. She'll be buried in the garden along with her predecessors.'

Lifting her in my arms, she felt solid and as heavy and unmanageable as a mattress; her tongue lolled sideways. I insisted we return her to the table while I tucked the tongue back into her mouth. The sight of it was too physical an image of death. Body fluids had begun to seep out of her and the heart specialist came running up behind me with half a dozen nappies. Suddenly, I was picturing Cleo that very morning, haring across the grass to accept the bone I had been offering her.

'No one's going to believe this,' I blubbered, half to myself. I could not believe it.

'Here,' said the dark-haired young woman, stuffing the towels between my arm and the dog's haunches. 'It'll help keep the car clean.'

The car! Did I care? It was a banger, a crock of a thing, practically held together with bits of string, used for the transportation of olives and the regular purchase of bags of cement required for the terrace walls that needed rebuilding after the wild boar had passed our way. A few abiotic drops, dog pee and dribble, would make precious little difference to its condition. I climbed into the driving seat once the rear had been carefully locked and started up the engine.

The windscreen wipers had jammed. Maybe they had been jammed for weeks, I did not know. The rain had only started to fall while I was inside. I gunned up the engine and set off slowly, climbing back up the hillside. In the late evening glow, Christmas lights glittered in the streets. Our village's central *place* was decorated with its annual, life-sized Father Christmas riding his sleigh drawn by four illuminated, bright blue reindeers. He was waving benignly at me, his clockwork hand tick-tocking back and forth. The bustle of traffic and shopping, the blur of raindrops sliding down the glass: all seemed like an out-of-focus film before my eyes, not reality. The reality was the deathly silence of such a pretty girl laid out behind me.

The following morning, after a desultory night, I awoke alone to sheeting rain. Michel had remained in Paris while lifeless Cleo had passed those final nocturnal hours in the rear of the car. Alone, I had not been able to lift her. In any case, I was not going to leave her body lying out in a storm. I hauled on old clothes and, half a cup of gulped coffee later, went downstairs. Quashia was looking for me.

'I have fed Lola and Homer,' he said, 'but I can't find Cleo anywhere. *Mon Dieu*, you look dreadful, Carol, what's happened? Have you been crying?'

I explained that the dog was in the car.

But why was the dog in the car?

I recounted briefly, avoiding details because I could not bear to repeat them, the fatal events of the previous evening. Quashia did not believe me, did not understand, did not want to understand, obdurately refusing to take on board the facts. This man had cared for all those tiny puppies alongside me and had stepped in with the milk rations on the occasions when Michel and I had been off on work trips. Empty as I felt, I had to spell it out to him, while we stood in the rain.

'I'll dig the grave,' was his bald response. He had no need to ask me where.

The lower of our two cherry trees, situated in one of the farm's ancient olive groves, had died and been chopped down the year before. We had planted a replacement close by the original's severed trunk.

Traditionally, it had been beneath that old fruit tree that all the farm dogs from our two decades of residence here had been laid to rest. The roll call had been standing at four: Sammy, Ella, Bassett and Lucky, who had been the most recent to be taken away at a venerable age. Cleo, way before her time, was to be number five.

Quashia bid me fetch his green jacket hanging up in the garage and then told me to get lost. He was intending to wrap the dog in the weather-beaten coat and bury her in it. I did as instructed and returned inside. Beyond the hiss of rain and the gurgles from the drains, I could hear the slap of the shovel against damp earth. I looked out of the window from my den and saw our man, standing in a muddied hole, pickaxing and digging. I poured him a cup of steaming hot coffee and returned outside.

'I want to do this alone,' he barked. 'I reared her. I'll see her into her resting place. Leave me be.'

His words were not entirely accurate – we had reared her, too – but I knew it was his method of dealing with his grief so I turned on my heel and left him to it. Swinging back, I called, 'Shout to me when you need help to carry her from the car. She's rather heavy.'

'How long have I run this place? And you think I haven't the strength to manage a dead beast?' He waved me impatiently away, determinedly concentrating on the earth.

Alongside the car, sitting alert, were Homer, Cleo's brother, and Lola, the running-to-fat, loyal mother; a cortège at the ready. Let no one claim that animals have no feelings. The expressions on their pointed faces left nothing unspoken.

An hour later, a knock on the door brought Quashia to me. His rheumy eyes looking directly at me.

'It's done?'

'It's done, and I'll be leaving shortly.'

'Yes, go back to the cottage and get dry. Have a hot shower. Thank you for what you have done for Cleo.' I was slipping my feet into wellingtons, intending to lay a flower on the freshly dug mound.

'We have no luck on this farm.' Quashia was now trailing me. 'This place carries bad luck with it.'

I rounded on him. 'That's not true and please don't say such a thing!'

'Yes, it is! She was barely more than a puppy. Last year at almost the same time Lucky died.'

'But Lucky was more than sixteen, Mr Quashia!'

'Bassett was poisoned while you were off doing whatever you were doing, away from here for months at a time, and we will have no olive crop next year, if your plans with that bloke go awry as I'm sure they will. As I said, no luck!'

We were back to the failed crops of former years. I sighed. Runnels of rain were dripping from my soaked hair, down my face and I knew that my instincts had been sound when I had suggested to Michel that we wait and not disclose to Quashia as yet our plans to finish with the insecticide sprays. It hadn't been necessary to say anything, but Quashia had been keen, nosing like a dog after truffles, to learn the identity of the visitor inspecting the olive trees alongside Michel.

'Mr Quashia, you have helped us bring in a select, hand-picked harvest of oil this year. Four pressings, all of which are top quality. There must be close to four hundred litres in the summer kitchen waiting to be decanted. I haven't even totalled it all up yet, but there is surely sufficient to keep us going for two years.'

'But we don't need to keep going for two years, Carol. Sell the oil, make some money. I know how hard it is to make ends meet here, so let's concentrate on upping our production level. I bumped into René at the food market a couple of days ago. He's got the right idea.'

'Michel and I both want to have another attempt next year at farming without chemicals, but, please, don't let us debate this again now. Not now, of all times.'

'You have tried those experiments before and we always lose the lot. We could have saved all those lost olives.'

'Mr Quashia, you are going back over history. As I have said, we have four hundred litres of oil down in—'

'Because we sprayed.'

'Please understand that we want to try and run this farm in a way that does not—' I sighed, exhausted. 'Can we discuss this another time, please?'

'But without the olives there is no farm, why can't you see that? I have decided to leave, Carol.'

'Leave, what do you mean?'

We were getting soaked and going round in circles.

'I'm going back to my wife and children in Algeria.'

'For Christmas? Yes, of course.' I knew that as a Muslim this was not his holiday period, but he had lived in France for so many years that he skipped between all festivals, using whichever best suited him for his and his family's agenda. 'That's fine, Mr Quashia, take a break. Michel and I will both be here from Boxing Day onwards and we can hold the fort. How long will you be away?'

He shook his head and stared at his feet.

'It's my teeth. They have to come out. The dentist said it could take months and months.'

'But you only have one tooth!' Actually, I bit my lip and just about refrained from remarking this fact.

'How long will it take to remove one tooth? Can't you see someone here? I can recommend a fellow down in Cannes.

As a matter of fact, he's originally from North Africa and his mother tongue is Arabic.'

'There are two at the back of my mouth . . .' Quashia opened wide, like a hippopotamus yawning, pink gums on display, and pointed to a pair of worn-down stubs, halfway down his throat, the colour of nicotine. 'They are giving me gip and they'll be as stubborn as hell to lift. I've had them all my life.'

'Must you go to Algeria? It seems a little far for one extraction.'

'Three. There are three to come out.'

'Yes, three. Excuse me.'

We were not getting any drier and, given his mood, his pain over our lost companion, I saw little chance of resolving the discussions there and then. I slipped my arm through the sodden crook of his and we set off slowly in the direction of Cleo's resting place. I was clutching the speckled white orchid I had clipped from one of the indoor plants. It now resembled a pad of soggy cotton wool.

'I expect to be gone six months. I'll leave early next year and I'll be back for the harvest, if there is one. You can telephone me at my son's *épicerie* when the fruits need picking. And if there's no harvest, fruits all rotten due to your new way of wanting to do things . . . well, there'll be little point in my returning.'

I took a deep breath and ignored this last threat. 'Six months, Mr Q?'

'And beside the urgent work needed on the roof in the garage, which by the way will bring the west side of the house down when it caves in, there's a leak sprung in the cottage.'

'Oh, Lord, I'm so sorry. Where is it?'

He did not reply. Wrapped within his silence my loyal steward plodded off, worn muddied boots squelching in the grey morning, heading down the drive to his cottage home without a glance back. 'This whole farm needs a firm hand,' I heard him mumble as he descended the winding path, disappearing from sight.

Six months! I hoped he was not serious. And this was the first mention of water infiltration down in the cottage. The whole place had been redecorated, its broken roof tiles replaced with new, only three years earlier. I gazed after him, not really focusing or even looking at the present, before stepping across the parking area, between the towering cypress trees to the lower of the cherries.

I was gambling that the loss of Cleo was the reason for Quashia's apparent despair and that we could help him sort out his three teeth before or after a well-earned break back in North Africa with his family. But I wasn't entirely convinced. I had noticed, slowly over the previous six months, since my return, an erosion, a gentle but decisive slipping away. Quashia had begun to withdraw. He spent less extra-curricular time with us. He rarely strolled up for a cup of tea on a Sunday afternoon, to talk through future projects, to chew the cud with his 'pal' Michel. Perhaps, instead, he spent his weekends sleeping, resting, watching Arabic television. He was getting older and was showing signs of slowing down, but who wasn't? I felt, though, that there was something else, as if a certain disapprobation towards us had crept in and was creating distance between us. Might it be connected to the extending arm of Islamic extremism in his country?

I wondered what, if any, inner turmoils were caused to those Algerians, village Berbers in Quashia's case and that of his pals, by the country's political and religious wranglings. This Berber man, originally a goatherd from the portals of the mountainous desert south of Constantine, whose father had been killed by the French during or before the Algerian War of Independence, had many split loyalties. We, our farm, were just part of the complex equation.

Quashia's wife, in a wheelchair pushed by our gardener's youngest son, her stepson, had made the Hajj a couple of years back. Our man himself had spoken of his desire to do the same this coming summer. Hajj. To pay his tribute, to undertake the requisite once-in-a-lifetime pilgrimage to Mecca demanded of all Muslims with sufficient funds to undertake the journey there and back. I wondered about the teeth. Six months? Perhaps it was his intention to fulfil his commitment to Mecca, to Allah, before it was too late for him, while he was still able-bodied and employed and had the financial wherewithal to pay for it. Possibly he preferred not to tell us about this. The teeth might well be his excuse.

I laid the soggy shreds of the orchid heads on Cleo's grave and stood for a while gazing at the sodden earth. Her mother and brother, who had been padding about behind me, came to settle at my side. The rain had eased and was falling softly now, tiptoeing, delicately dancing off the sods of turned earth as though marking respect.

Gravesides. Burials in the rain, they emit such a mood of gloom, don't they? Despondency. I turned and trudged back to the house. On my way, I stopped by the garage. In his hasty retreat Quashia had forgotten to lock the door. As I pulled it to, I noticed that the floor of the interior was a shallow lake of water and yet more tiles had fallen from the ceiling and lay broken within it. Raindrops were dripping fast on to our collection of tools and machinery, sliding down their plastic coverings. My management efforts had been in vain.

It was almost Christmas.

In a few days, I would be leaving for Paris. Michel had organised for us to spend a little of the early festive season up north with his twin daughters and their burgeoning families; travelling was easier, less complicated and costly for us than for all of them. I was pleased about this. I loved the young life around us, the gaiety and energy and the pleasure it gave my husband. Clarisse had moved far from us into the higher Alps, to a remote skiing village that was not the easiest to access. Michel had been trying to contact her, without luck, and we had not heard a word in more than two weeks. Her baby had been due the first week of December. It was looking unlikely that she would be meeting us in Paris.

Standing in the rain now, tugging my feet with difficulty out of my green wellingtons, I felt a ghastly sense of gloom, a prescience that was nagging at me. Were there further losses to come? Would Quashia genuinely consider leaving us?

Returning? What does that mean, to return? How do you return? Can you retrace your steps, wander blithely up past the gate lodge, stroke the dogs, take a tour of the groves and settle yourself back into your habitual chair? One of the certainties about travelling and then returning is that while you have been absent, engaging yourself in a million new experiences, the life you left behind you has also moved on. Everything evolves, changes, grows or diminishes. My life and that of the tribe surrounding me was no exception.

I was chucking woolly clothes, long socks, boots into a suitcase, collecting fresh herbs from the garden for the seasoning of the Christmas Eve turkey. Layering lemons and bitter oranges direct from the trees into shopping bags. These citrus globes served several purposes, culinary, comestible and aromatic; quartered and stuffed into fowl or served as dressing with fish, sliced into fizzy water or gin drinks, or peeled skins placed on burning logs in the fireplace to scent the rooms. Musky Arabia. I needed reading material to laze by those fires; books, writing pads. All was in anticipation of my departure to Paris. At the same time, I was also preparing for our return. This entailed shopping, stocking the cupboards and fridges with extra wine and non-perishables, and generally organising the rooms so that we would land back at home to a warm house and convivial atmosphere in readiness for guests arriving for the end-of-year celebrations.

Quashia's mood had improved just a little and so had mine, much supported by a reappearance of winter sunshine, and, most importantly, the joyful tidings that Clarisse had given birth to a girl. Hitting the scales at five kilos, this little bundle was to be christened Chiara, which, of course, translated as 'light'. Perfect for the season and to illuminate and elevate the mood I had been in danger of sinking into. Delightfully, for Michel, who had not seen his girls or their offspring for several months, it was agreed that everyone would convene at Vanessa's in Paris, including Clarisse accompanied by her brand new infant, but not, alas, Chiara's father, Philippe. He ran a skiing resort and this was his high season. She was to stay with us for a couple of days in the creaky old house Michel had moved into during our time of separation a while back now. This property outside Paris had since become the closest we had to a base in the north, essential for his work, which was predominantly centred in the capital.

Vanessa, the second of Michel's twin girls along with her husband, Cole, had recently relocated to the outskirts of Paris into a more spacious first-floor flat. They had moved to this new address a few months earlier and neither Michel nor I had so far found an opportunity to visit. Previously, the couple had been existing in a one-bedroom, top-floor walk-up in a lovely part of the heart of the city, but the place with its ancient facilities had offered no lift and Vanessa, who had recently and most unexpectedly found herself the mother to three, had been obliged to drag prams, shopping baskets as well as all the rest of the child-rearing gallimaufry, up six flights of winding and narrow eighteenth-century stairs. It is rare, I understand, for the twin gene to be directly passed on and appear in the succeeding generation, but that was what had happened in this instance. Vanessa, a twin herself, had given birth to twins. She and Cole, who were not in a financial situation to extend their family by another two, had accepted their lot with good grace – they both adored their babies, of course – but I had been aware of the strain on Vanessa. She had relinquished her job as a photographer, a profession she had loved and had been extremely gifted at, and had decided to return to university to qualify as a teacher. She was starting again from scratch but she was tenacious and dedicated. Her new role would entitle her to a secure salary and longer, paid holidays. Time that could be designated to the raising of her children.

During the period Michel and I had been separated, a lonely, isolated time for me, I had seen less of his daughters. Understandably, their devotions had been torn or, rather, not torn at all: their loyalties had sat firmly in their father's camp. Once or twice, at larger clan gatherings – funerals, birthday parties – it had grown awkward and I had stepped back. I had sadly felt obliged to let them, as well as their father, go. Fortunately, that was all a long time ago now and the restitching of relationships had been successfully achieved. Still, I was aware that there were gaps. I had missed out on certain developments within the family, the children growing up, and, because everyone was constantly occupied – all running about doing their best to make life happen, to earn a living – some of those gaps had never been closed.

Marley, Vanessa's oldest boy, was a point in question. Marley, who charged to and fro like a fireman on a hazardous mission, was approaching eight and was six years older than his sibling twins. Vanessa had been pregnant with him when she had stood at the floor-to-ceiling window in her office in Manhattan and had watched the destruction of the Twin Towers in 2001, had witnessed the jumping of victims from windows to certain death. Had anything of the complex emotions of that horrific day and its aftermath been transmitted to that unborn child forming in her womb?

Michel and I, though our relationship had been strained at that stage, were there together at Roissy-Charles de Gaulle airport a year or so after 9/11 waiting to meet and greet Marley upon his first arrival into France, a sleeping baby who had crossed the Atlantic to begin his new life and had known not a thing about the devastation of the world at large, the shock ripples that were still permeating. Vanessa, with her American husband, Cole, armed with their small child, were choosing to build their future life in Paris. New York, they had decided, was, at that stage, no place to raise a family.

They had been laden with luggage and I had been handed Marley, a fair-haired, chubby cherub heavier than half a crate of olives. Looking back now, I think it was not an accident that Vanessa had passed her son over to me. She had always been sensitive to my situation. I stared hard at the boy, his eyes closed, long blond lashes, as though attempting to divine him, to know the secrets of his future and his sleeping heart. Impossible, of course. But I did mutter to Michel later when the parents were elsewhere that I wondered whether the boy might be carrying any traces of the memory of America's recent traumas in his psyche.

It was Christmas Day, mid-afternoon. Along with Clarisse and Chiara who were staying with us, we were congregating at Vanessa and Cole's for champagne *en famille*.

Their twins, a boy and a girl, had recently turned two. Marley, as was his habit, was leaping about the place, calling, whirligigging, entertaining, wearing a selection of hats, wielding a flashing sword, brandishing a spaceship that was emitting an invasive low-frequency sound. The centre of attention though, whether Marley cared for it or not, was Chiara, who slept serenely, burping and gurgling her way through the festivities. For a child that had weighed in so solidly at birth both Michel and I had been surprised to be introduced to such a delicate creature, but, boy, was she gorgeous. A veritable

Sleeping Beauty. Clarisse's eldest daughter, Anna, was absent from these celebrations. She had opted for spending a week with her father, Clarisse's ex-husband and a long-standing work colleague of Michel's. Among this intimate gathering was one other grandparent, the children's blood grandmother, Michel's first wife. Some might consider such a set-up a little unusual, but this was France. Sylvie and I were friends. I admired her and enjoyed her wit and humour. She was pretty and she was a caring and attentive mother and grandmother. And we were not in competition. She had her own life, another relationship, lived for her offspring, was neither career-obsessed like me nor always on the move. When we arrived she was sitting in a corner of the living room playing with the twins while Marley twirled like a top and played the fool, no one throwing him more than the occasional glance.

As a family, we sat drinking tea and later champagne, talking of those inconsequentials that fill out such occasions, that crochet together the stretch of time since we had all last assembled like this. Marley was designated the task of passing round the nibbles, which he did relentlessly and seemed aggrieved, disappointed, if we did not accept yet three more cheese-flavoured biscuits or another supermarket olive. A general and genuine interest in my travels provoked questions. I recounted anecdotes and spoke of future plans for my Olive Heritage Trail, my contacts with UNESCO who were keen to work with me on the project, and then the conversation returned to subjects of daily life while Cole and I talked theatre in a corner. Since moving to Paris, he had begun to direct stage productions though it was not easy for him to find employment, he confided. In these matters, within the family circle, I was his obvious point of contact. In fact, we had worked together recording the English narration for one or two documentary films produced by Michel so we shared professional bonds.

When the general conversation turned to the hurdles of schooling, education and finding a decent crèche, I was always glad to share Cole's company. These were debates where I held no opinions, had no experience and I never involved myself in them. This was a delicate issue for me. Occasionally during such moments I suffered a minor sense of exclusion, but reason prevailed. Why would anyone discuss such affairs with me? These were not my subjects. I was childless.

Marley was tugging at my cardigan, determined to draw my attention away from his father. Cole and I were conversing in English so he, a bilingual boy, followed suit. He was trying to coax me to follow him into his bedroom. He wanted to show me 'a secret'. I laughed, winked at Cole, excused myself and tramped along the freshly painted corridor behind the marching blond-haired child, madly waving his upper torso about as though he were a tree caught in a tempest. He closed the door after us and set to flicking on lights and beepers and then began yelling. He dragged from beneath his bunk bed a small trunk, flew it open and disgorged on to the ground an assortment of toys, books and plastic warfare components. He had switched to French now and was bidding me look at this and that. I felt uncomfortable, uncertain what to do.

'Is this your secret trunk, Marley?' I was reminding him of why he had beckoned me here, for he had possibly forgotten the pretext already.

'A secret, your secret?'

He paused for a moment, a much needed bridge of stillness, and then he sat back on himself on the floor and stared at me as though he had never set eyes on me before. Immediately, he leapt to his feet, flung the door open and screamed for 'Pappy-Michel' and 'Maman'. I feared I had upset him, or had broken the spell in some way by asking a question that was too mundane or logical. Vanessa came hurrying along the corridor and grinned at the mess he had created. Slowly, quietly, she began to switch off the flashing, bleeping hardware and restored the room to a semblance of quietude. And then she smiled that electric smile of hers that could emblazon a coalmine.

'Great flat,' I said. Marley was now hugging my legs. 'I love the colours, the soft olive green and cream of the woodwork.'

'And I painted the entire place myself,' she laughed. It was an open laugh, a sound that tossed her head back almost recklessly, swung her shoulder-length dark hair from side to side and left her milky neck open and vulnerable. Her beauty was breathtaking. She was such a sensuous woman that I sometimes silently bemoaned the fact that she had gone for motherhood and marriage so readily, at such an early age. I had always pictured her in some international arena, but such a thought voiced would have simply amused her, caused her to laugh more loudly still.

'Your vision of the world is not mine, Carol,' she would have said.

And yet.

'Well, it's thanks to this little lad here that we have such a great place,' she said, ruffling his hair, which seemed to subdue and calm him now.

'Really, why?'

'He's a special needs child. It gave us preferential treatment on the waiting list for a bigger apartment.' Vanessa spoke English with a broad American accent.

'Special needs?' I repeated, completely taken aback.

'Yes, Marley's hyperactive personality disorder, ADHD, plus the fact that I gave birth to twins gave us an opportunity. I could not have stood traipsing up and down the stairs of that Paris apartment any longer. I was studying for my teacher's exams – phew, it was getting tough with two small ones and this one bustling about the place as though he were about to explode. Cole, when he is rehearsing, is working every hour under the sun. Well, you know how it is. I was going insane.'

I looked at her hard. Her beauty was paler and she was strained, lined, but less so perhaps than the last time I had seen her.

She and I had been close at one time. I adored her and after she had moved to the States, before she had met Cole, she had written me a wonderful letter that I had kept, treasured. She had told me that I had been 'a role model' for her. I was living a woman's life, her prose had informed me. Spending time with me during her formative teenage years had allowed her to see that there were possibilities out there and it had encouraged her to be brave and set off alone for New York, to complete her master's there. There, where she had met Cole. I had read the words, the sentiments of that letter, over and over to myself on many occasions, particularly during the period of marital separation. It came flooding back to me now.

I had seen so little of her. For a number of years. We had drifted apart and yet whenever we were together I felt a great rush of love, of admiration, a childish desire to claim her as mine, my flesh. My daughter.

And all this while she had been struggling.

'You must come and stay with us,' I whispered. 'Bring the children and have a break.'

'Heaven knows when,' she grinned, 'and where would we put the babies and this crazy little fella?' but she leaned in to me and wrapped her arms about me. I could have cried.

Fortunately, at that moment Pappy-Michel arrived and Marley was up and running, keen to display the gadgets again. Vanessa left us with him and returned to the main room, to her mother and the rest of her dearest kin.

'I had no idea,' I whispered to Michel in English.

'About what?'

'Marley and his special needs.'

Marley glanced my way, eyes narrowed. He must have understood.

'It's really very minor. He has no physical handicap nor any mental disability. It seems to be psychological, poor attention span, I think, but it has given them certain privileges which they were very grateful for. They struggle to make ends meet and they could not afford anything larger in Paris. This is ideal for them.'

'Why didn't you tell me all this?'

Michel was crouched on his haunches, constructing some sort of magic castle with a dragon hovering at its moat. Boy and man were on the ground pulling bricks and plastic sticks of furniture from out of the trunk. Marley watched his Pappy silently, entranced, following his lead. 'I honestly didn't think about it, *chérie*,' my husband said to me. 'I assumed that you and Vanessa would have discussed such matters between you. I know how close you are.'

'No,' I replied absent-mindedly. I was fascinated to note how calm Marley had become, how absorbed he was in the world his grandfather was creating.

'Is he a dragon, Pappy, or a snake with wings?'

Michel is quite wonderful with small children. He draws them into extraordinary imaginative universes, but what was fascinating me was how readily Marley had responded.

In the world of theatre I have known many actors, musicians, dancers and artists who were cast aside at school because they were judged to be incapable of concentration, difficult to teach, unmanageable, when, in fact, it transpired later that their talents lay in other directions. They were artists. I confess to being a little wary of such diagnoses as ADHD, Attention Deficit Hyperactivity Disorder. I myself as a child growing up in a home where there had been deep-scarring traumas had been frequently chastised because I was 'incapable of sitting still'. I was not incapable at all. I was anxious to absent myself, to transport myself into the lands of make-believe where I felt safe and out of danger, where my eyes could look upon other, less distraught relationships, and

I was at my most contented and peaceful when I was alone with the fantasy tools of creativity. Looking at Marley now with his large, round eyes locked upon the busy hands of his grandfather building a tower, I wondered.

4

Luke returned to visit us early in the New Year, with an exercise book scored with figures, estimations, flight path capabilities for the fly, timetables of cargo boats from Africa. It was impressive, well studied. Luke's complex calculations suggested that we would require, exclusively for the lower groves where the old, gibbous fellows resided and our main fruit production took place, one thousand flies. He was offering to sell us five *paniers*, baskets, each containing two hundred of these exotics. Each batch equally divided, he said, between one hundred males and one hundred females.

'And this will cost?'

His response to Michel did not sound exorbitant. 'I can ship them, get them through customs and quarantine and I'll include delivery of them to your farm and even assist with the releasing of them in May; no, on second thoughts, perhaps we'll do it much earlier. Yes, earlier. Depends when I can land them here. So for the entire service, I'll charge you four hundred euros a container. In other words, two grand all told.'

Obviously, we did not write him a cheque there and then. In fact, from the beginning, Luke made it perfectly clear that he would be requiring the money in cash because his 'contacts in Africa insisted upon it'.

'I see.'

'It's delicate,' he said.

We should have heard the warning bells, foreseen what lay ahead right then. My only excuse was that I was rather too eager to make this work. And I had done a fair amount of homework, researching *Psyttalia*. *Psyttalia lounsburyi*, as well as similar parasitoid relatives of his, were being reared and studied under quarantine in research laboratories in Sophia Antipolis, not fifteen kilometres from our farm. Infested olives had been gathered and taken to the laboratories for use in their experiments. The fly was also in residence at other research centres in California, Hawaii and southern Italy. Judging by a photograph I found of him, he was a rather attractive predator, minuscule, with a pair of outstanding, disproportionately long antennae. I surfed the internet day and night, but most of the information was hard to access and appeared to be under wraps. Still, the more I learned, the more excited I became because – and this was the hook – *Psyttalia lounsburyi* promised to be a serious – and organic – contender for the role of Olive Fly Exterminator.

Michel, on the other hand, was growing concerned. He, who had initially been enthusiastic and supportive of our move away from insecticides, felt that we were jumping the gun. It was not that he suspected Luke, any more than I did. It was that Michel doubted the wisdom of setting loose an alien, a creature untried and untested on our land.

'This could be irresponsible,' he confided one evening in mid-January.

'Oh?'

'If the laboratories are not releasing the insect, it is because they are not yet convinced that it is safe for the environment.'

⋆

By the time the first month of the year was closing down, the weather had turned incessantly wet. Rain greeted every dark morning as I rose early. Michel had flown to China on a business trip. I was alone at the farm with Quashia. Together, we stared into the space that had once been the garage and was now a lake of detritus.

'I warned you and I'm glad I listened to me and not you and moved all the essentials up into the *hangar*. If you don't repair this roof, the west wall of the house will crumble beneath it. It's too wet to prune or to reconstruct the dry-stoners. If you want me, I'll be in the greenhouse.' And with that he was gone, my loyal follower.

The fact of the matter was that we could not afford the repairs on the roof. The quotations had put paid to all our plans at this stage.

Luke dropped by. 'Listen, I've got a flight booked to Nairobi on Friday. A rather substantial order to fulfil. Are you in or not?' He was advising a swift decision. He felt that the flies would do better to sojourn the latter part of winter, early spring, on our land to acclimatise themselves to our situation and the environment.

'Is that what your other clients have decided? I'd love to be in contact with them. Always heartening to meet others who are fighting for the organic route.'

'They're in Italy. Rather private people. It's not my place to give out their details.'

'No, of course not. Look, I'd prefer to wait until Michel returns . . .'

'I think it's urgent that you make the choice, Carol, so that I can get this order in, get the little sweethearts on the boat and on their way here. There are limited numbers of these darlings for sale. I don't want to lose this shipment.'

'Of course, I understand. Can I let you know definitely one way or another tomorrow?'

I could not reach Michel. We were operating on a six-hour

time difference. Whenever I telephoned his hotel room, he was not there or sleeping.

Vanessa called. She was trying to reach 'Papa', wanting to let him know what a pleasure it had been to spend time with him at Christmas.

'Why don't you come down for a weekend, bring Marley?' I offered. 'I'll get the plane tickets.'

'Hey, we'd love to but there's just so much on. Tell Papa, hi from me.'

'Yes, of course.'

That night alone in the house, music playing softly, working into the small hours at my computer, I came across a series of documents that led me to independent scientific research findings drawn up by Greenpeace. The results comprised a blacklist of agrochemical products hazardous for human health. Dimethoate, our olive spray, was on the list.

The rain had stopped and Quashia was itching to commence the pruning. The small trees had never been cut back and they had grown tall. They were lofty but without sufficient width on their lower trunks to secure them. He warned that a tempest could bring them down. His agitation heightened my dilemma, but I cannot lay the blame at his door, not at all.

On the Thursday morning, after another call from Luke who warned that if we said no now, it could well mean a year's delay to our plans, I took it upon myself to make the decision. I accepted Luke's proposal, drove down to the bank and handed over two thousand euros to him in cash, but I was careful to take it from my own pool of funds rather than the farm's. Two euros a fly. Was that an extortionate price to pay for what I hoped was a clean future? I had no idea, no notion of how to appraise such a costing. These little beings were arriving from Africa, after all. And if Michel was not in accord with my decision, I would argue (rather feebly) that it was my money I had lost. Still, I handed it over with a degree of incertitude. These two thousand euros, I said to myself, might be used towards restoring the garage or alleviating a little of Vanessa's constraints.

'*Madame*?' The voice at the other end of the line reminded me of Marlene Dietrich.

'Yes?'

'It's José. I have found someone who can handle your asbestos roof.'

'Big José?' I cried with delight. 'Good to hear from you.'

There was a silence. 'No, Madame, I am rather small. Are you still looking?'

'Yes, we are!'

'I will bring him to take a look.'

'And you recommend him?'

'He'll get that asbestos shifted for you at no vast cost and little hassle.'

'Excellent.' I did not admit it but we had been almost desperate enough to shell out the seven thousand euros we had been quoted at the outset.

The following day, José, the shorter, wirier of the quartet, arrived in his van followed by a gentleman on a motorbike with a clipboard. I thought at first that his colleague, excessively tall, built awkwardly, like a misshapen barrel on sticks, was German. The man was introduced to me as the asbestos expert. He was employed by one of the companies who had refused me, but who were willing to do the business if I went through José's masonry firm.

'Before you begin to dismantle anything here, you will need a certificate,' he informed me, looking at the few sheets of roofing and shaking his head despairingly.

'I thought the certificate came last,' I argued. My instinct was against this fellow from the outset. 'I understood we received that when the material had been treated and disposed of.'

'I am not talking about *that* certificate. I am referring to the other one required by law to begin the work.'

This was new information. 'I am going to call the *notaire*,' I retorted.

'Madame, we need to prove beyond doubt that you have asbestos here. If this roof has not been fabricated from a dangerous substance you can bury the sheets in the garden and say no more about it.'

I considered for a moment, trying to read between the lines. Was he offering *at a price* to deny, nullify the official registration of asbestos, discreetly recommending that I deal with this toxic nuisance illegally?

'My job is to verify one hundred per cent that this material has not been wrongly diagnosed because there has been no laboratory testing of it.'

'Yes, there has,' I insisted.

'No, I would know it if there had been. You require a registered certificate and this might very well prove that there is no asbestos here. Then, the good news is that you are off the hook.'

How could he know whether or not there had been a lab test?

'Sir,' I countered. 'The reason we are in this extremely aggravating situation is because an expert, such as your good self, came here and after laboratory tests declared that this roof is one hundred per cent asbestos sheeting.'

'Who was this expert?'

I marched into my office, dug out the house files, riffled through a mountain of papers and furnished him with the woman's name.

'I knew her husband, a decent fellow. Poor woman has been down on her luck since his death. May I see her report, please?'

I unclipped the twelve pages and handed them over to him. He took them with hands that were red and puffy, fingers like rhubarb sticks.

'As I suspected, no laboratory test.'

'Of course there was a test! Look, here.'

He shook a big shaggy head that reminded me of a punctured football. 'I need to verify this report. It has not followed the precise letters, codings of the law. We need to prove that a laboratory test really did take place.'

'Monsieur, if you are offering to dig off another morsel of this unattractive and undesired corrugated covering, to "test" it again and furnish me with a negative report, thus clearing my liability, then please say so.'

'Good Lord, Madame, I could never do such a thing! That would be a criminal act!'

'In that case, sir, I am content to go along with this lady's findings and follow the legislation. Can your company, please, give me a quote for the removal and destruction of this toxic material or not?'

'Not without a report from me. It is my written certificate that counts.'

'Then I am sorry that you have been troubled and that you have made this visit for nothing. I wish you a good day, sir.' And with that I turned on my heel, throwing a daggers glance at the Portuguese mason.

But this visit had fired my determination to find a solution once and for all to the dilemma. I returned to the Yellow Pages and spent the latter part of the morning ringing every listed builders' yard. From there, I contacted construction engineers, demolition firms and one company who dealt in dangerous materials and, they, the penultimate call, led me to a building expert in Grasse whose operator put me through to their technician. I explained the crisis and its urgency.

'Well, we certainly handle asbestos.'

'Can you send someone to give me a quote, please?'

'What for?'

This stopped me in my tracks. 'The asbestos. To know what it will cost us to have these sheets taken away, deconstructed, disposed of.'

'The rates are standard. I'll fax them through to you now.'

And he actually did. The rate was priced by the weight of the sheets, wrapped and prepared for departure by the ton. Compared to the months of aggravation we had suffered, this seemed to be remarkably simple and very reasonable. Including VAT, it worked out at approximately five hundred euros a ton, which, compared to our quote of seven thousand, had to be a gift. Unless there was a catch somewhere. Might those few sheets exceed any poundage I could possibly imagine? Fourteen tons of asbestos equalled seven thousand euros. Impossible. I telephoned Michel, now back in Paris. How much do those corrugated sheets weigh? I begged.

'*Chérie*, I'm in a production meeting.'

'Yes, of course, sorry.'

Quashia had no idea of their weight either. I called through to the technician in Grasse and accepted his offer.

'Now,' he said, 'a cautionary word. The quote represents the cost of the removal and the legal destruction of the material.'

'I see . . . What else is there then?'

'Either you, your builders or someone from my team will have to remove the sheets from their current position and then stack them, and they must be stacked squarely one on top of another.'

'Fine . . .'

'And then, and this is absolutely an imperative, they must be wrapped in a specific, listed, plastic sheeting. It is of a designated density and protects against toxic leaks. If you cover the stack in anything less substantial, less impermeable, the driver will refuse to take it. And, by the way, he will under no circumstances touch the stack with any part of his person.

All contact with the package will be achieved by machine. I will fax you the details of the plastic covering and its dimensions. Nothing else, and I mean nothing else, will be accepted. And we make no reimbursements. If the driver goes away empty-handed you are still liable for all charges.'

I felt a rising sense of uncertainty.

'Any chance that you could supply the plastic sheets for us?'

'I'm sorry, we don't offer that service.'

'Well, where can we purchase them? At our local builders' merchant?'

'Possibly. Let me send you the details. Three other points: I will need a five hundred euro deposit and we will need to fix the pick-up rendezvous as soon as you have faxed back the signed copy of the contract, agreeing our terms. There is a small fee for changing it, once confirmed. Last point: when the driver has removed the sealed package, he will give you a paper. This must not be lost. On no account. It confirms his receipt of the material to be freighted and it is the copy of a form that will be given to the toxic waste disposal unit who will sign it and send it back to you once the material has been destroyed. Both will need to be forwarded to your *notaire* who will then file them at the public records office. Is all that understood?'

The builders' yard did not have the specified model of sheeting or rather they did, but the thickness and dimensions were marginally different and I was way too unnerved to take any risks. They knew me well at the yard, because I am there at least once a month buying sacks of cement and ordering deliveries of sand to repair our damaged walls. I explained my situation and I begged their help.

'Oh, Madame Drinkwater, poor you. Let me put you through to the Hairless Goat. Just one moment.'

The Hairless Goat?!

'You're through to Mark, how may I help you?'

I was dying to ask Mark why he was known as the 'Hairless Goat', but refrained. I poured forth yet again my requirements and he calmed my anxieties by assuring me that 'it would be a breeze'. I would have his response within the next couple of days.

I faxed back the contract to the disposal firm, posted them a cheque for the deposit and felt thoroughly pleased with myself for having finally overcome what had threatened to be an insurmountable hurdle.

Michel was equally delighted when I told him later.

'How is the production going?' I asked.

Exhausting. He was taking the evening off to babysit for Vanessa.

Invite them for Easter, I suggested.

By the time Michel had returned to Europe, spent some days on business in Paris and flown down to the farm, Luke had niftily disappeared into thin air. His answering machine remained active, but he did not respond to any of my messages. I had no address for him, no other details.

'He must still be in Africa – that would explain his non-communication – organising the cargo,' I argued to Michel who said nothing, clearly displeased by the action I had taken.

But the days rolled over and the silence grew louder.

I was beginning to feel perfectly stupid. I did not own up to what had happened to Monsieur Quashia. He would have laughed me off the face of the earth. Instead, I excused the delay, the lack of arrival of promised flies by fabricating a half-truth: further research, study was required . . .

'A fly that eats a fly and solves our harvest problems. I knew it was ridiculous.'

'But we would still like to steam ahead with our plans, Mr Quashia, to convert to an organic holding for this upcoming summer. If *Psyttalia* does not work out, I have been reading about a plant, a chrysanthemum. I am having difficulties finding information about it and I am not yet sure whether it is available in France—'

'Don't tell me, Carol, an African insect-eating plant!' Quashia's mocking words, but at least he had not quit. In fact, the removal of his teeth seemed to have been quietly set aside. However, if I was to keep faith with him, guard a semblance of respect, as well as maintaining his enthusiasm and commitment, I had to deliver an alternative solution, and swiftly. To defend my position, my fervent desire to live on a farm that produced only organic foods and protected us and the environment, I needed to understand the facts better and come up with a viable alternative to the sprays I was so adamantly condemning. If we lost yet another crop, our manager and loyal friend would withdraw all confidence in me. His threat of leaving would become a reality.

There were no messages from Mark, the Hairless Goat, on my mobile so at the end of the afternoon I went through to my den to check what might have come through and I realised that I had overlooked the cheque. It was still in the photocopy machine along with their fax. I had posted the signed contract without the payment. I picked up the phone and tried to reach the *société* in Grasse who had agreed to dispose of the *amiante*, but they were closed for the day. It would have to wait. Tomorrow, when they received my envelope, I would explain.

I called the builders' yard and asked to be connected to Mark. 'There's no Mark working here!' a girl at the other end of the line informed me.

'But I was talking to him this morning . . .?'

'No, sorry.'

'Well, could I speak to the, erm, Hairless Goat, please?'

'Oh, him! He doesn't work here. You would have been linked through to his mobile. He's not here. He's the Hairless Goat. Hang on.'

Which I did. And, while doing so, I tried to picture this bald apparition who had earned himself such an extraordinary nickname.

'His mobile's not responding. I'll send an email to him to call you straight back.'

When Mark telephoned, the news was not good. He could provide the material, the sheeting, but the *épaisseur*, its concentration, was insufficient, less than had been instructed.

'Oh, I cannot possibly take it then. They are exceedingly strict.'

'Well, there's one place that does supply that size, but they charge one thousand euros a sheet and it sounds as though you'll be needing three or four sheets.'

'What?!'

He gave me their name. It was the company which had given us the quote for seven thousand euros. They seemed to have this particular market sewn up, I was thinking while pouring out my heart to the bald fellow at the other end of the line. 'Mark, I'm running around in circles here. What can I do?'

'That's why you're talking to me. That is what the Hairless Goat is here for. Why don't you just double-wrap whatever the material is?'

'I can't. It's asbestos!' I spoke the word as though it were an atomic weapon.

I could hear his grin. 'Just double-wrap it. Or triple it, that'll be perfectly fine. It's the same difference!'

'I'll come back to you.'

I left an urgent message on the personal mobile of the technician who was handling the removal, apologising at the same time for the fact that his accounts department was about to receive the original signed contract without the cheque, which was now in another envelope and would be on its way in the morning. I begged to know whether I could use a thinner version of the same sheeting and triple-wrap it. I was certain the response would be negative.

He called back while I was cooking supper to confirm that he would not cancel the appointment. Fine for the delayed cheque and, of course, I could triple-wrap the plastic. It was no problem at all.

So disappointed was Quashia in my lack of capability, my vision as to how this holding of ours should be run, including the lack of hives, that he took to hiding the garden tools. I would go in search of a pair of secateurs, to prune a branch, pick off the heads of some herbs in the greenhouse, cut a few daffodils for the table, no matter what, and they, along with my leather gloves, maddeningly, were nowhere to be found. Because I am always obsessed with owning an example of every variety of whatever it happened to be, we were the proprietors of seven pairs of secateurs, as well as, between the three of us, eleven pairs of gardening gloves including leather, cotton and rubber prototypes, plus a wide assortment of boots: wellingtons, sneakers, sandals, walkers, Tasmanian Blundstones (given to me on a work trip Down Under and treasured by me as about the most comfortable footwear on God's earth). All went missing.

'Monsieur Quashia!' I would yell from one end of the property to the other. The onset of his deafness aided his stubborn refusal to pay me or my needs any attention. 'Where is the . . .?'

When I eventually unearthed him buried away in some corner or other, he would simply shrug as though he had no knowledge or interest in the matter. 'But you must know where they are!' I insisted, working myself up into a lather.

'I do know. They have been put aside, concealed.'

'But why, *why* have you hidden them? I need my boots and the pruning shears now, Monsieur Quashia. Where are they, please?'

'I am not giving them to you.'

'What do you mean you are not giving them to me? They

are MINE, Monsieur Quashia, MINE,' I would yell at him, exasperated, tears of perspiration jumping from my forehead.

'You don't know what you are doing and you will turn this entire property into unproductive dead wood, if you get your way. I see it as my duty to protect the farm.'

'Protect? In what way protect?'

'Against you.'

This hurt and I could see from his steely expression that he was not fooling. 'I am doing my best, Mr Q. You are unaware of the dangers.'

'No, Carol, you are stubborn and you are listening to other people's nonsense. I have been working the land since I was ten years old: building stone huts, pruning olive trees, herding sheep. I was born among olive trees. I know what I am talking about, and you are lost.'

'But the world has changed,' I reasoned.

He shook his head, as if to say 'I am closing my ears to you'.

'Man is destroying the earth, putting it and us at risk and we must learn to respect it. Care for water, the soil . . . for example, no more burning plastic in the garden or throwing non-biodegradable substances over the fences on to ground that does not even belong to us . . .' (Two other causes of dissension between us.)

'It's only rubbish. No one will find it.'

'But that is not the point, Mr Q. It is not good for the earth.'

'For heaven's sake, Carol, farming our olives is what is good for this earth. And while I've got your attention, what are we going to do about the bees?'

'I have explained to you what happened.'

'Well, why don't you find an alternative solution? Let's have our own hives. I'll take charge of them.'

'It's a skill, it has to be acquired and neither Michel nor I could—'

'You find the bees and I'll care for them. You know how

I love honey. When I was a boy in the mountains, I used to go off with my brothers and friends, a whole troop of us scaling the mountain faces seeking out the nests of wild swarms. And when we found them, we'd steal the honey. It was risky, oh boy, but once we'd filched it, triumphantly we'd stuff ourselves with it there and then or we'd carry it home with us in goatskin satchels and then store the honey in clay jars. Those illicit hauls were the most delicious honey I have ever tasted. It's why I'm always telling you that supermarket rubbish is not even the same product.'

'Yes, but it involves more than just helping yourself to the honey,' I cautiously suggested.

'I know that. I've been here with François and watched him working with the hives and I want to be a beekeeper. Can we count on the return of a few hives? It would make me very happy.'

'I'll do my best.'

'Aside from olive oil, you can't do better in life than have your own honey.'

'You're probably right.'

And with that he would slope off, and still I would have to go in search of whatever had been sequestered.

So, we had a firm who were willing at a reasonable cost to transport this infernal roof and relegate it to its final end, but we still had no one to achieve its removal. Michel suggested that we manage it ourselves. Lift off the sheets, package them and store them out of harm's way until the pick-up truck arrived to cart them off to their last port of call.

I scooted to the builders' yard to buy the requisite plastic sheets and pick up some protection masks. At the counter, when I gave my name and order number, the female assistant said Mark wanted to explain precisely how the enfolding should happen.

'I thought he didn't work here?'

'He's here today.'

I waited. Into the storeroom walked a young muscular man, mid-thirties, with a healthy mop of shiny brown hair.

'Hi, I'm Mark. The protective film is out in the yard. Follow me. We'll cut it to order for you. Quite a hoop you've been through, eh?' He grinned. We grabbed a packet of masks and, puzzled, I followed him out into the goods yard. He cut me four generous sheets for good measure and we returned inside to the cash desk.

'Anything else?'

I shook my head, still wondering about him. 'One question, perhaps, if it's not too indiscreet.' Michel, if he had been at my side, would have chastised me.

'Sure, fire away.'

'Why are you called the Hairless Goat?'

Mark threw his head back and his Adam's apple rippled with delight. 'Ha! That's a good one.' He swung round to one of the men working behind the orders desk. 'She thought I was the Hairless Goat!'

'Ha! Ha! Pierre, did you hear that? Well, some might think of you with certain tendencies associated with that old beast . . .' The word had spread. A mixture of merriment and lewdness had ignited the routine of the morning.

'When did you last see a hairless goat?' Mark asked me, eyes glinting, still enjoying the joke.

'Never.'

'Precisely.'

My goods had been parcelled up and Mark handed me my bill. A grand total of fifty-seven euros.

'A little better than seven grand, eh? We aim to please.'

'I'm sorry to be stupid, but I don't get the joke.'

'My job,' he said, spelling it out to this dimwit, 'is to find the impossible for our customers. If you need or want a hairless goat, we'll find it for you. *I* am not the Hairless Goat. The department is. We serve our best and most long-standing customers. We make sure we find them everything they need, whatever it is, no matter how difficult or impossible. So now you know where to find me, next time you're stuck. Good luck with the asbestos.'

He was walking me to my car, carrying the heavy package.

'Interesting, the song and dance the authorities make about this material, isn't it? Yet, years ago, when folks like my dad were working on the sites and people shouted that asbestos was toxic and was damaging their health, no one listened. I knew loads of blokes when I was a kid who got sick from being around it, inhaling it, ended up with lung cancers. The authorities tried to cover it all up, until it all came out and they were forced to admit the facts.'

'Yes,' I nodded, opening the car, shaking his hand and thanking him for his help. 'I think it's rather similar to other issues we are fighting today.'

The nuisance roof measured a mere five metres by three and a half. It had been battened down with nothing more durable than a series of securing bolts, which Quashia and Michel managed to release without any difficulties whatsoever. It was a bright, warm Saturday morning and once the sheets were untethered we lifted them away with caution. It was vital that they did not crack or crumble. The dust particles from the asbestos were the danger, I had been warned by Mark. We managed the process between us without snapping any of them. The men then lugged the sheets across the parking and placed them, one neatly on top of another, on to a wooden palette beneath which four layers of plastic grey sheeting lay, in readiness, once the stacking had been completed, for the water-tight sealing of the shipment.

'Look at this!'

I was still over by the stables staring into a gulley within the brickwork, laid bare by the removal of the roof. There, a thousand lives had been exposed, rendered vulnerable and homeless. Geckos, scurrying in every direction. I had frequently noticed them breasting the walls, pausing cautiously mid-step, when I had been inside this space loading or unloading the washing machine, but I had never been aware that here thrived an entire community. Secreted within these cracks and crevices were several dozen individual nests, each with three or four eggs, round, small, white, perfectly formed and without markings. The parents had fled. Families broken. At first, I thought that all the eggs had been hatched but on closer inspection we found dozens that had not yet opened. It stopped us in our work.

When Quashia went off to his cottage to grab a spot of refreshment and Michel was up in the greenhouse cutting us a salad, I hurried into my den and dug out an animal encyclopaedia. Ours were Mediterranean House Geckos, *Hemidactylus turcicus*. Over lunch on the terrace, I suggested to Michel that, if it were at all possible, we might try to incubate the eggs of these orphaned creatures.

'I wouldn't have a clue how to go about it, but I agree it's a pity to let them die if there's a way to save them.'

The gecko holds a very special place in the Olive Farm hierarchy. It was not unusual for these reptiles to make their homes within cracks, fissures, substrate of human habitations, and man and these squat little lizards with their splayed suctioning feet are able to live together in perfect harmony. It is a mutually beneficial relationship. Geckos feed off insects and we have always been delighted to have them trekking across the ceilings, targeting the mosquitoes, padding back and forth – slow, slow, quick, quick, slow – or bedding down throughout the daylight hours in the cupped wall lights. Since our earliest days here, when we had not a centime between us and slept on a lumpy mattress on the floor in front of the fire in the *grand salon*, I would watch them, clamping their finger pads efficiently against the chimney plasterwork as they shuttled up and down its breast. I listened to their squeaking.

Such a force of sound emanating from a being so small. Back then, our very first winter here, I pronounced them our lucky mascots and, ever since, there had never been a time when I have not encountered them secreted between walls and open shutters, keeping cool in the shade.

On many occasions, I have almost trodden on one of their minuscule babies, the texture and shade of worn rubber or plasticene, lurking within the furling corners of our antique Turkish rug, skedaddling for dear life, fearing the fall of my giant's step. But what I had never come across before was an egg, a breeding site. This heretofore invisible space between corrugated roof and plastered walls constructed out of aerated bricks was the hub of their colony, judging by the number of unhatched eggs.

While Michel and Quashia secured the asbestos, I did a little research, hoping to learn how to hatch the unborn creatures.

The Mediterranean gecko is a reptile native to southern Europe and, interestingly, is believed by scientists to be resistant to pesticides. Clearly, it was not endangered by the toxicity of asbestos! I read up on incubation tanks, correct temperatures, but to my unscientific mind it seemed rather complicated and I was uncertain whether I was really capable of orchestrating the birth of hundreds of geckos! I returned outside to count the eggs. There were one hundred and eleven, white and round as peppermints. Was this a feasible exercise? Delicately, with enormous care, in between fingers, I picked one up, like a mothball, but it instantly disintegrated and out of it fell dust.

With every single egg I lifted from its carefully hidden cache, the same story. They crumbled to sandy dust.

'And to my amazement,' I was recounting my discoveries to Michel over lunch, 'I've unearthed internet clubs, societies of people all over the world, who breed geckos. I watched a couple of short videos on how to care for them when they are sick, but none explained the sawdust substance.'

'You know who would have been fascinated by your find,' said Michel when I told him. 'Marley. When I was babysitting him recently, I noticed his fascination for dragons, dinosaurs, the mystical lives of creatures. Next time I have a free afternoon in Paris I'm going to take him to the Natural History Museum.'

After lunch, when Quashia came up with the post he handed me just one letter. It had arrived from an agricultural syndicate whose function had never been entirely clear to me though, for over a decade, since we had registered ourselves as olive farmers, we had been paying them bi-annual union fees. After several visits from one or other of their experts, we had been admitted to this mutuality. Allegiance, membership, was obligatory if we were to remain eligible for our hard-earned AOC. Their missive stated in two very bald sentences that due to our lack of olive crop the season before last (their books always seemed to be about one year behind real time), therefore no oil and, more importantly, no litres put forward for an annual quality-control rating, we had been dismissed from the association. No warning. No 'regretfully' . . . Simply 'Disbarred'.

Michel was not pleased. He asked me to file this communication with its companions and he went off about other farm business. His parting words were 'it took a great deal of time and effort to get ourselves accepted by that crowd. If one doesn't play the game, Carol, you get nowhere in France.'

And that was the crux of this issue. As far as everyone was concerned, I was not playing by the rules. The Luke episode was the perfect example. I had attempted to circumvent the system and it had backfired. I called his number one final time and found the line had been disconnected. His story must have been a fabrication.

Organic farming is a lonely road. It is becoming less so but for the time being it belongs to the 2 per centers: mavericks, visionaries, those who are not only passionate about the virtues of their food but also about the future of the planet. Organic farms represent a mere 2 per cent of all French agriculture. Even given that this country is the biggest producer and exporter in Europe, that is a very small fraction. Organic *olive* farming was a lonelier path still, because there were no solutions. It was not simply more difficult and painstaking, in our coastal region of France it was virtually impossible and fraught with obstacles; the olive fly was resistant to everything except chemicals. But I remained determined to find a solution. My hopes were now pinned on the efficacy of the chrysanthemum though I had still not traced it. Of the five garden centres I patronised, not one had heard of it and no one could furnish me with its name. Handsome Alexandre though, who ran the farmer's cooperative I frequented, a passionate hunter and acquaintance of ours, was able to tell me that, although he did not know anything about it, he had heard talk of this flower and he believed it was endemic to the Dalmatian coast. He promised to find out some facts for me.

'Still chasing the elusive, are you, Carol,' he winked as I stepped into our van. Such a flirt he was.

'The elusive?'

'Just spray the trees, sweetheart.'

Daffodils were pushing up through the newly laid paving stones down around the cottage. The bell birds were in fine throttle. We had been dreaming of achieving the first stage of this extension before summer, as an incentive for the girls and their babies to holiday with us. The days were marching on. Spring would soon be on its way. In the hope that it would considerably reduce the exorbitant quotes we had been receiving, when Michel was home at the weekends he and Quashia began digging out the area beyond our bedroom, above the garage, foundations for the new roof spanning garage and stables on to which a fine terrace was to be laid.

Still without hopes of construction afoot, we were now

living with a trench, expanding by the week to its full seventy-five metres, right outside the bedroom doors. Our two stables, where the dogs had slept and laundry room was still situated, were roofless. Michel rigged up a temporary cover for the washing machine area, the dogs were given a warm, dry home up in the *hangar*, but the garage was beyond all solutions. Every time I entered there, another portion of ceiling hit the deck, frequently damaging whatever remained beneath. Holes were being revealed in the rear wall where plumbing and electricity pipes protruded from every crack and crevice. It was a disaster.

Our aspirations lay with the Portuguese and, while Michel haggled over prices, I concentrated on the olives. The Chambre d'Agriculture represented the conventional guys, the official bodies, and I felt beholden to touch base with them. I wanted to find out how they were getting on with the battle against Dacus and to learn of any new developments in the world of olive farming. Might they have come across the Dalmatian plant I was searching for? How were they responding to scientific reports such as that published by Greenpeace? Due to all my travels, it had been close to two years since I had last contacted them and I was now informed that the splendid young official we had always dealt with in the past had left the organisation. In his place was a female technician, Maude, who invited us along for a day of *formation*. I smiled, picturing all those knotty, dyed-in-the-wool Provençal farmers who were now obliged to deal with a woman, and, what was even more challenging, a woman in her mid-twenties, judging by her delicately light voice.

'We are keen to work with farmers such as yourselves,' she explained on the telephone, 'those who are not full-time agriculturalists, but who see olive farming as more than just a hobby and who are committed to producing fine table fruits and oil on an annual basis.'

Michel and I agreed to attend their one-day seminar due

to take place the following Saturday, though I found it a bit curious that we should be perceived as part-timers given that we were registered, proven olive growers and had been designated an AOC for oil excellence.

'You don't seem to have understood, Carol. It's because we have been struck off their lists,' emphasised Michel. 'No doubt, they will re-enlist us at some stage, but it will be a slow-moving ship and, undoubtedly, we will be expected to jump through the same bureaucratic hoops all over again.'

'But why? We produced a perfectly decent harvest and pressing this last winter and it will be on file up at the mill where they are obliged to declare their daily returns.'

'Those figures will not be in the system until next autumn, when all the papers have been filed. We are not allowed to register them before May, as you know. For at least the next year we are off the agricultural radar screen and that's a fact. Our farm does not exist as far as any of these organisations are concerned and it is up to us to turn that around, again.'

The first part of the day's tutoring was held at an agricultural research centre built within allotment lands on the banks of the Var river, nestling in between hills and overlooking a vast industrialised plain behind Nice. It was the tail end of February and the sun had broken through. Signs of early spring, although the morning lectures were taking place indoors in a makeshift lecture unit. These were to be dedicated to the olive farmers' fight against *les ravageurs*, the insects and fungi that played havoc with the trees themselves as well as the crops. The position of our property, so close to the sea with a south-westerly aspect, was in many ways a blessing because it provided us with full sunshine from an hour or so after sunrise all the way through to sunset and offered us plenty of air and light, distanced as we were from mass construction and far removed from any other farm. The downside was that the fly hit us first. The warmer it was, with plenty of saline humidity, was just how Dacus preferred it. The little beast thrived in such climates. Only those holdings set at altitudes above six hundred metres were free of his antics. And with climate turbulence on the increase, those establishments might also soon be threatened. Meanwhile, we were right at the top of his hit list. Almost no other olive farm in south-eastern France sat as close to the sea as ours did.

Since the unfortunate incident with Luke, I had been trying to nail down a contact at INRA, the National Institute for Agricultural Research. Somewhere behind their closed doors, the fly *Psyttalia lounsburyi* was residing in quarantine and I wanted to know how realistic was the possibility that he would soon be at liberty and out in the fields. I was hoping this information would be available to us on this Saturday morning. I did not doubt that this exotic would be on the agenda.

The lectures were being conducted by Maude, who was indeed in her early twenties, and her colleague, a slender dark-haired woman of similar age. Both were employed by conventional agricultural organisations under the auspices of the Chambre d'Agriculture. There was no voice, no method proposing the organic route. The products on display were those that had been on the market for a number of years. The only lip service paid to the growing demands for a greener alternative was the recommendation to the farmers in the audience to show greater caution when dealing with pesticides containing high levels of toxicity. Slowly, these were to be withdrawn, both women informed us.

But when and to be replaced by what?

Until the previous year, certain of the products had been enthusiastically recommended, including dimethoate (I never received a response from their manufacturers in Belgium), and we olive farmers had been actively encouraged to spray our crops five times throughout the summer months, the season of fruit maturation. Diazinon was another product in question (also on Greenpeace's blacklist). Up until the previous year it had been recommended by the Chambre d'Agriculture, then farmers were requested to limit spraying to once a season. Now Diazinon, as a pest control, was being phased out of many international markets. Outlawed, marked as 'entirely unsafe'. It decimated bird flocks (no mention of human risk). We ourselves had used it several years back, until, after a little home research, I had rejected it.

So, at what stage, I was silently asking myself, do we stop listening to these people? With certain products such as our dimethoate-based chemical, a maximum of three times a year was still permissible. ('You sprayed with this product five times last year, and it was deemed safe,' I whispered to Michel.) And next year, or the year after, when they have discovered some horrific side effect, or are publicly admitting what they must already know, what will they propose then? Was not asbestos a clear example of such mismanagement?

Not one single item on the agenda for that morning was now legalised for five sprayings in a season.

Products, including dimethoate, were still on the market, still available, but we were advised to use them 'with greater caution', to exhibit awareness of their negative properties. I recalled the proposals set out in the previous November's manifesto by the Minister of Agriculture: to create awareness among farmers. Was this the sum total of what had been intended by that declared goal?

The second compromise of the morning: it was now illegal to harvest crops during the twenty-one days that followed a spraying. In other words, farmers were obliged to hold off gathering their fruits until the toxic residues in the olives had diminished.

But it had always been twenty-one days, I grumbled under my breath.

We were also advised by these two young female technicians that the use of products should become less habitual, if possible. 'Don't go for the gun, as a matter of course. Think first, do I need to kill? Use only for remedial purposes rather than preventive.'

'And how do we do that?' called someone behind me.

Hang traps from the trees to catch the flies. This was to give an indication as to the density of the infestation. Once the farmer was sure that the flies were present and propagating, soon to be injecting their eggs into the maturing olives, then they were at liberty to spray.

The traps, though, were not specific; they trapped whatever wandered into their zone, they killed other insects including honeybees.

I sat in my chair in the front row next to Michel, fuming. Michel glanced my way and winked. He knew me too well. But I was not going to take up the sword. Not here, or not yet.

I glanced behind me at the rows of faces, a gathering of some seventy or eighty of us in total; worn-down, drink-sodden were some; rusted by the elements were others while a few wore skins as wrinkled and runkled as old geese. Typical land folk, but what surprised me was the expressions they wore. Fed up, bored. Evidently we were not the only pair who had hoped to hear more inspiring news, guidance towards a greener future.

A man in the row behind us to the right with apple-pink cheeks and a fraying straw hat, a figure of bucolic bliss, raised his hand. 'Aren't we supposed to be trying to move away from all these chemicals? They're downright toxic,' he called out. 'What organic routes are being studied?'

Good fellow!

'Yes' was bellowed from a big-bosomed woman to the left, shuffling and scuffling her brown clods of feet. 'We know all this.'

'We are coming right up to that,' replied one of the female technicians.

Then, before she could inhale and continue, someone called out from the back row. His words were distorted by an echo within the tinny space, but I clearly caught the gist of the question: what are the latest developments on *Psyttalia lounsburyi*? So, the exotic fly with his long and rather attractive antennae was a well-known player within the olive community, a highly anticipated star turn.

Our hostess sighed as though this was a question that weighed heavily upon her. 'Nothing' was her response. 'He is a failure.'

'What?!'

'Not possible!'

'Are you sure?'

'I cannot believe that!'

She lifted her hands to hush the rising clamour of contentions. 'As many of you will know, this exotic, this African fly, has been imported and is being studied at government funded science laboratories in Sophia Antipolis. It is a five-year project. Well, *Psyttalia* survived into its second year and before Christmas was taken out into a pre-selected list of groves for a trial run.'

'And the results, what were the results?' Impatience from a heavy woman in thick-laced shoes.

'The result was nothing.'

'Nothing?'

I had proposed parcels of our farm, a selection of both young and elderly groves, for these experiments, but my tender was received too late. The five olive locations from this region had been selected while I was far away on my travels and I had failed to stake the opportunity for us. When I learned about it, I was disappointed. I had been keen to observe the experiments at first hand, though at that stage I had not known what the experiments were to be. It would have been a gamble.

The previous November, batches of the African flies had

been transported to the pre-selected olive fields, we learned now, hung within unsuspecting branches and left to their own devices. The flies were to make their own way, to find their own place within the groves' ecosystems, but, according to this technician, the insects had refused to leave their boxes. A few ventured out a yard or two but no further. Three weeks before this day's meeting, as the first signs of winter were receding and creeping out from beneath the earth came the tips of the wings of spring, a group of scientists had returned to see how the little foreigners were coming along. However, every site had registered the same disappointing message. All flies had disappeared. There was not a trace of any of them.

Voices were calling. What had happened?

These particular colonies had been reared in incubators under highly cosseted conditions, their origins were hotter climates . . .

And it had been an unnaturally damp winter along this coast of the South of France. Rains had veiled the views, had reduced the crumbly red soil to a streaming burgundy soup. Leaves had dripped, boughs had hung heavily, some had broken beneath the onslaught of water.

'And the flies?'

'Yes, what of the flies?'

'They just perished, were washed away. Consequently, the *Psyttalia lounsburyi* exotic has been deemed unsuitable for our more northerly temperatures. As a biological control agent against the olive *mouche*, it appears that the propagation of this species lacks efficacy.'

Sighs and groans of disappointment from the audience. I threw a swift glance about the room. Clearly, we had not been alone in putting our aspirations into this little fellow who was no more than half the size of his combatant, Dacus, who himself was half the size of a fingernail.

'Any questions?'

I raised my hand. 'I have read about a plant that might have originated along the Dalmatian coast. I believe it was, is, related to the chrysanthemum, contained insecticidal properties and was used as a natural insect repellant in the Middle East and possibly the ancient world, too. Unfortunately, I have not been able to track down its name. I wonder if anyone here is acquainted with it and knows whether or not it might still be available.'

Bemused faces greeted me. I turned to our seated colleagues. Perhaps one or two of those who leaned towards organic had Googled the same results as I had, but the room remained silent.

'Sorry, Madame, *désolé*. However, we have some good news to report: a relatively new pest control, developed in the United States. Insect management is its key function. It has been created out of the fermented residue of a plant and it carries an organic certificate.'

Several members of this farming fraternity brightened up. 'That's more like it!' 'Let's hear the good news instead of all this doom and gloom.'

In some detail, Maude's colleague, Irène, described the product, and I recognised its name. I had read about it during my late-night internet trawlings. It had been patented and was being sold by a major American pesticide company, a company with a dubious history, and this pesticide was highly toxic to bees. When the young woman had finished and asked if there were any questions, I raised my hand again.

'It is true,' Maude's colleague admitted, 'that there is a minimum risk for bees but as honeybees do not pollinate olive flowers, they are not attracted to the groves. Honeybees rarely frequent olive groves, therefore the dangers for them are slight.'

Now I was ready to leap from my seat. Michel slid a hand over my left thigh to temper my reaction. It was unnecessary. The man in the straw hat with the ripe cheeks was on his feet, dancing with anger, defending the world's most vital pollinator, the honeybee, *Apis mellifera*.

'All bees are endangered! It has been publicly acknowledged the dire predicament they are in. You, as representatives for the Chambre d'Agriculture, have no business condoning any chemical that eliminates even a tiny percentage of the world's remaining arthropods,' he was calling, his cheeks rouging up, shinier than before. Mumbles of agreement hummed in the cavernous space.

How, I asked myself, had this product received an organic ticket when bees, no matter how few, were endangered?

Our afternoon passed in less contentious circumstances and in a far more agreeable environment. After a picnic lunch in lukewarm sunshine at a long wooden table outside the tin edifice, the two girls walked us up the hillside to the higher ridges of a low mountain. It was quite a hike. We were now in the company of an ample, middle-aged man with cheerfully blue, piggy eyes who hummed contentedly, like a distant diesel engine, throughout the climb. We entered what until recently had been a vast swathe of mimosa forest that descended sharply towards flatlands. The golden mimosas had been dug up, the hillside replanted, converted into extensive groves of young *cailletiers*, olives. These junior silvers had been arranged in rows, but somehow they were not regimented, more welcoming, and they were of an astonishing height, thrusting themselves upwards out of this steep incline. Their bushy crowns overlooked the industrialised plain of St Laurent du Var and beyond to the sparkling Mediterranean. The holding's aspect was full sun. This afternoon it beamed out of a soft, liquid azure sky that seemed to rise upwards from the horizon's rim, arching above us and then curving backwards until its blueness slipped discreetly into the depths of the beyond. Looking about me, it seemed that the whole world was a blessing, a globular expanse of sky and silver vegetation, a vegetal viaticum.

On this beautiful Saturday afternoon, beneath such a compelling heaven, the trees were swaying gracefully, touched by a whisper of wind beckoning spring. It was hard to believe these saplings were just five years old as our expert – square fingers, big square nails – who spoke with a thick Provençal accent, informed us. Our youngsters, six when we purchased them, had been in the earth for more or less the same length of time, perhaps even a tad longer, but they had not bolted, had not grown so lofty or bushy as these examples, even though we also had never pruned ours.

By now, still humming, our good-natured agriculturalist had stepped into his blue serge overalls, pockets bulging with the tools of his trade: secateurs, a small handsaw, an electric branch-cutter attached to a portable battery, a demi-litre of mineral water jutting out of his breast pocket. At his side stood an assistant, his daughter, who resembled him in every way with her porcine features and squat, thickset physique.

'We are here to learn the unique and essential skills of pruning saplings. Impossible, of course, in one afternoon!'

Standing in a circle round this father and daughter team, who had travelled from their birthplace, a forgotten hamlet within the high, blustery rock of the Massif of Sainte-Baume (*baoumo* in Provençal means cave), especially for this demonstration, all of us were perched, feet entrenched firmly into the ground, at a most precarious angle. Monsieur le Tree-Cutter began to lift branches, tendrils, to examine them, and then with a magician's sleight of hand he flicked lengths of bough, of twig or entire towering leafy limbs off the trees, quizzing us, in continuous dialogue with us, our thoughts on which strand to extricate next.

'From where should the following snip be taken and for what reason?'

I noticed that encircling the foot of each tree was an arid patch of earth, barren, where no herbage grew. I guessed the owner of the estate used a weedkiller around the plants' trunks to save himself a chore. All the while our woodman was working, he conducted conversational exchanges. He described the structure of the olive tree, its organism, its ability to moderate its perspiration levels, thus making it highly drought-resistant, its process of internal communication, from root to leaves.

'This living being here at my side,' he smiled, 'has, in my humble opinion, a far more efficient central control system than, let us say, a Boeing. It relays messages from its roots to leaf tips, advising the leaf cuticle when to stop perspiring, sending warning signals when the water levels below ground are insufficient and the plant cannot afford to lose moisture, when and to what degree it needs to shift into its protection-against-drought mode. Its water-conduction system is first rate. As is its capacity for survival. We have all heard of computer-sophisticated aircraft alerting pilots to the fact that fuel levels are dangerously low, but are those mechanisms capable of cutting all engines and gliding the machine home? I doubt it, but this tree can regulate its own moisture system . . . it can hold its liquor,' he laughed, 'and it's never going to nose-dive and kill all relying on it!'

He spoke of the advantages of lifting off canopy height and of leaving the lower, outer branches to tumble and swing. It is said that a healthy olive tree is one that a swallow can fly through without the tips of its wings brushing the foliage. A bird's width between branches allowed the interior of the tree's crown to be aired, which in turn lessened the possibility of fungal infections, keeping undesired humidity at bay. A well-pruned tree has the ability to combat disease.

Unfortunately, it does not protect against the olive fly.

'So, besides chemicals what solutions are on offer to us?' I asked.

'Well, to start with, keep the entire farm healthy. A well-maintained grove has a higher level of resistance against infections and a better chance of combatting infestations. At the end of the harvesting season, clean the ground of all residues of olives. They are the winter homes for the flies. Leave no unpicked fruits on the trees. Rake them up, burn them.'

After a little more delicate snipping, our professional paused, raised his head skywards, squinted, studied the tree appraisingly and announced that he would cut nothing more from it.

'But the work is not finished!' cried one onlooker.

'I have trodden gently,' he replied. 'I have pared, rather than lopped. These are not my trees,' he explained, 'and I have never worked on this hillside before. So, this fellow and I have not yet become fully acquainted. I have not yet had an opportunity to understand the needs of this tree nor has he got to know me. We are strangers,' he continued, with a shy smile. 'We have not learned the language of one another, how to communicate our requirements to one another. Next year, if I were here again, it would already be easier between us.'

He spoke as though this was a love story, or at least an intimate, personal relationship, one broad arm held high, moving in curves as though caressing an imaginary outer silhouette of the plant. Our instructor pointed out the new growth and drew his thick stubby fingers, thick as broad beans, nails ingrained with soil, along the slender, curving roddy extensions where fruiting had taken place the previous season and he showed us where not to cut if we wished our juniors to deliver us oil this coming autumn.

'We are in the presence of an organism of infinite wisdom,' he grinned. 'But don't let that panic you. If you prune inaccurately, these guys are tenacious. They are not going to allow you to kill them off. A few bursts of energy and they'll reshoot, but while they are readjusting, finding a new equilibrium for themselves, you are usually deprived of their bumper crops.'

While others threw questions at him – 'Why not take that bit out?', 'Those shoots could go' – or showed off their own expertise, I pondered this country steward's words; with his paunch, his fully fleshed face and dense southern accent, he spoke with poetry of his vocation, his relationship to the tree. He spoke as though these trees were human or he was fluent in their language. I was delighted to hear a grown man discoursing with the assumption that such a language existed. But then his ancestors would possibly have spoken such a tongue. His birthplace, the region of Sainte-Baume, constituted massive and quite spectacular forestlands. During the centuries of the Gauls, it was honoured as a sacred forest and remains a protected site even today. He would have been raised on country lore that paid its dues to trees.

How would it be, I wondered, if we could interpret the trees' concerns? If the earth could talk to us? But then again, I reflected, it *is* expressing itself and we are not listening.

5

The asbestos was to be removed. Hallelujah!

At a quarter to two, fifteen minutes before the material was booked to be taken away, I received a call from a driver sitting outside in the lane.

'I can't get this truck up that drive, not even through those gateposts,' he spat down the phone.

I had explicitly requested a van that could pass through our narrower-than-average entrance.

'Well, no one told me. I work for an independent firm, love. We're haulage contractors and no one said anything. Hang on a second.'

A few minutes later a portly Arab came huffing up the drive. He was muttering under his breath, curses to Allah. He had left the two blue gates open and I could see that parked beyond was a mammoth carrier used by furniture removal companies.

'No, you'll never manoeuvre that in here,' I confirmed.

'I know that! Have you got any staff?'

'Our gardener.'

'Tell him to carry the parcel down and we'll load it down in the lane.'

'But it weighs a ton!' Quite literally. 'Perhaps more. He cannot possibly carry it single-handedly, nor even with your assistance—'

Hands held up as though in defence. 'I don't touch this stuff, Madame. It's toxic and my contract states—'

'Yes, yes. Well, it will not go into the rear of the farm car or I'd ferry it down myself.'

'I'm leaving.' He turned to go. 'I should have been forewarned.'

'No, wait, please. I'm sure we can find a solution.'

Quashia must have heard voices because he came strolling down the hill to greet us.

'*Bonjour, jeune homme*,' he called, as he approached, which is his salutation to all visitors no matter their age or status. As soon as the two men set eyes on one another, they slapped their hands together in greeting and switched languages. They were now conversing in a mixture of Arabic and a Berber dialect.

Whatever spell Quashia weaved I had not understood, but the fellow nodded, turned, set off down the drive and waved.

'Do you know him, Mr Q?'

'No. But he's an Algerian.'

In this country, that was a claim to brotherhood.

'So, what is going to happen?'

'He's bringing up a loading trolley. We'll wheel it down between us.'

'But how will you lift such a charge on to it? He said he won't touch it and, unfortunately, I don't have the strength to do it. You'll break your back if—'

'Hush, Carol, we'll manage.'

As his fellow countryman reascended our steep incline, I was reminded of those airport luggage carts that simply will not obey. This mechanism was far lower, twelve centimetres off the ground, which meant that perhaps we – they – could probably slide the package on to it, but its wheels were behaving as though each had been given a different instruction. The good news was that this fellow Berber's humour had been radically revised and he had stopped complaining. Thank heavens. We needed the signed form he was going to hand over when this burden was finally off our property.

It took all the strength of the pair of them, the driver with thick gloves, pushing and pulling, yelling, bawling instructions, to heave the asbestos on to the wheelie. Once achieved, they had to negotiate it down the drive. Its weight and wheels gave it the advantage. It took control. I had returned to a spot of chair-painting when I heard hollering and ran to take a look. The two men were literally being dragged down the hillside, zigzag fashion, as this mobile devil, with a frenzied life of its own, took off, careening towards an ancient olive tree. I pictured the smash, the asbestos within breaking up. Toxicity! Quashia put on some speed, champion that he is, and overtook the blasted pushcart. I feared it would splay him on the tarmac but somehow, don't ask me how, he managed to break its impulsion. It spun almost in a circle as it slowed and Mr Q was hanging on for dear life, rather like one of those cartoons where someone grips a moving object with their hands while their legs are outstretched, four feet off the ground. The other man had let go altogether and only caught up with Quashia three-quarters of the way down the hill where the incline eased up. They were the masters of it now and slowly delivered it beyond the gates. At the rear of the van was a hydraulic lift to raise and disburden it within. The job was done. Those corrugated sheets were our responsibility no longer. We, the three of us, climbed to the table beneath the magnolia tree. I offered tea of freshly picked mint leaves, which brought smiles to their sweating faces. The chauffeur scribbled in the details of what he described as 'This'll be someone else's Christmas present now', drank his tisane, shook hands with our hero and handed me the pre-certificate, proof that the collection had finally been accomplished.

'You are a star, Monsieur Quashia. Thank you.'

'Well, if you left everything to me, Carol, this place would run like clockwork. And what's the news on the bees you promised me?'

After our excellent lesson from the tree surgeon the weekend before I was firing on all cylinders, raring to try my hand at sculpting our young beauties. I had no intention of attacking the ancients. Ascending a ladder, perched on high with a whirring chainsaw between my fingers was too daunting a prospect for me. Quashia and Michel could handle the collosi between them. I was perfectly happy to tame the entire younger brigade, if necessary, an ensemble of well over two hundred, but as soon as our gardener saw me ferreting about in the *hangar*, gathering utensils well hidden, unpacking boxes he had secreted there himself, he was after me, eager to know what I was about.

'The young groves have never been pruned. It's high time for their first cut,' I announced.

He agreed.

I was also, a little cunningly, I confess, calculating that if all the trees on the estate were lopped hard, while remaining faithful to their graceful, billowing form and the structure our tree-cutter had shown us, then, while regaining ground, they would produce less fruits this upcoming summer. Less fruit would obviously mean that there would be a reduced harvest. If we stayed on course and kept the land free of chemicals, the trees, the fruits, would be unprotected. If they were preyed upon by flies, the likelihood was that we would lose the crop, just as we had done two autumns earlier. However, the loss would be easier to accept because fewer fruits had been produced to harvest.

'Let's go,' Quashia said, grabbing both large and small chainsaws and various other hacking instruments that he kept buried in the boiler room, well out of everyone else's reach. I explained patiently that he could continue with whatever he had been doing because I was intending to manage the younger groves myself. His initial doubt as to whether I was capable of the task vanquished, he suggested that while I was within range we should deal with the eucalyptus trees.

On Appassionata's hillside we have two tender-blue-leaved eucalyptus trees. Both were growing close to the house; both had been left to their own devices and soared vertically, but Quashia had a point: they lacked definition and were spindly.

'They need trimming right back.' He was pointing to their higher branches. 'We should shorten them, chop them to the central trunk and then they'll grow strong, less willowy.'

'No, they're fine. Eucalypts can grow to heights of up to one hundred metres, depending on the species, Monsieur Quashia. You cannot treat this genus as though it were an olive, an *oleae*. And, frankly, you gave them quite a haircut last summer before the party. I think it's best to leave them.'

'They'll rejuvenate.'

I preferred not to get caught in a debate. The day was calling. I had been considering taking the two eucalypts out altogether because their roots were voracious; they drained underground water sources and left surrounding plants in difficulties. I must have muttered my thoughts aloud because Quashia hotly disagreed. 'We can't lose them, we need them.'

And he had a point. The other side of the coin was that the eucalypts were a rich source of pollen for honeybees as well as other pollinators.

'You're right, Mr Q. We will keep them and we'll just gently prune them, but not today, eh? A nip and tuck later, nothing more.'

He seemed convinced by my argument. I was delighted and set off to begin the awesome task of pruning two hundred and thirty young *oliviers*.

'Bees, Carol.' Quashia called after me. 'Give me hives! What are you waiting for?'

That evening I telephoned François and learned that he and Marie-Gabrielle were still attempting to sell all their materials associated with their honey business, but they had as yet received no serious offers.

'People are nervous about the investment. News of the bee crises is spreading. There is little confidence left in it as an industry.'

Still, they had managed to secure a bank loan and had begun works on the conversion of two newly conceived chalet rooms. The *gîte* would supplement their pensions. It was a heart-rending situation for them, for bees, for the state of the world. Quashia had a point; we should find an alternative solution and so I proposed purchasing their hives, taking all their apiary equipment off their hands and funding new colonies, if they would agree to go into partnership with us.

'We can keep everything down here on a permanent basis,' I said, offering François the role of Olive Farm Beekeeper if Qashia could be his assistant and I might play the part of second assistant. This arrangement would involve François driving from his mountain home once every two or three weeks to check up on the girls, housekeep the hives, teach our gardener.

'Quashia is all for the plan and I miss the bees hopelessly,' I added.

It was the truth. I longed to sit crosslegged on the grass,

gazing at them. Hours I used to while away, entranced by their comings and goings. I found it therapeutic, meditative and I was never once stung. They were usually harmless and gentle and they delivered us delicious honey. In the spirit of business, I suggested to our friend that we split whatever honey was produced on a fifty-fifty basis.

I detected from his responses that he was tempted. However, I also detected a new strain, a thinning in the timbre of his voice and a breathlessness. It suggested to me that his health was deteriorating. His words revealed disillusionment; he sounded aged. Still, I was hoping that such a proposition, one that released him of all financial risk, would reinvigorate him, offer him a new lease of life as well as providing us with the company of bees again, but I was honest and admitted that my one nagging concern, the worry that had been staying my hand, was the chemical spray.

'I do intend to put a stop to it, though,' I promised.

'But what alternatives do you have, Carol? There is nothing else out there.'

'Nothing concrete yet,' was my answer, 'but I am determined to find a way forward.'

François promised to mull it over and get back to me.

After months of toing and froing, the Portuguese eventually undercut Bolmusso's quote by more than 50 per cent, bringing the overall figure in at a snip over thirty thousand plus VAT on condition that a reasonable percentage of the total was handed over in undeclared cash and the rest was drawn up as restoration work.

'But it is restoration,' I insisted.

'Ah, but there's to be an extension.'

'Yes, but not on this contract. We haven't applied for planning permission for that section yet.'

Still they were insisting on 60 per cent in cash. And, in return, 'We will agree not to describe the works on the final statement as a "new extension to the garage", but simply as "repairs to the older part of the house". That'll lessen the final figure.'

This differentiation meant that the overall total qualified for lower VAT rates. Five point five per cent instead of nineteen point six. Older properties benefited in such cases. It had been an initiative by one or other of the French governments to boost the number of restorations on innumerable listed or historically interesting properties rather than encouraging the French habit of rejecting the old and constantly throwing up new chunks of brickwork.

The finer details of the sums to be paid in cash and what was to be declared proved to be so complicated and time-consuming, delaying the decision, that Michel proposed he draw up a plan and they thrash it out when the men had set to work.

'Any idea how long the job might take?' I thought we should know.

'Four weeks, maximum five. It's not complicated.'

'When can you start?' I asked, when everyone had shaken hands on a marginally illegal arrangement.

'Monday,' replied the short one with the lustrously husky voice.

'Goodness, so soon. Well, it needs to be done, so, all the better.'

Marie-Gabrielle, instrumental in her husband's decision, I felt sure, returned my call early the following morning.

'I think we must refuse the offer, Carol. My love is getting on,' she confided to me, 'and his heart has been broken by the loss of so many girls. I fear he has not the strength to go through it all again. You know, he has suffered a minor stroke since I spoke to you last, once he fully comprehended that we had lost everything. It wasn't about the money, well, partially, yes, of course – both of us had invested our life's savings. It was the destruction of his dream. All his life he had planned for such a retirement: our chalet in the mountains and a hundred hives.'

I graciously accepted their refusal, of course, but I was disappointed and at a loss as to where we might find a replacement for our previous arrangement. On a human level, I was saddened by the tragedy our friends were enduring. Quashia was also not delighted when I relayed the news to him.

'Well, if they don't want to come back, why don't we buy our own hives? I'll look after them. We can place them over on the Second Plot, out of harm's way, behind the oaks, backing on to the extended wall I have rebuilt. It's a dead space. We have nothing growing there but weeds.'

He had a point. It was a leafy and protected position.

'Look at this.' I had found a grey caterpillar. It was attached to the olive branch I had just severed. I lifted the insect from the tree, turning it over; its soft underbelly was a light silvery blue. Its upper body was identical to the colouring of the tree's wood.

'What kind of caterpillar is this?'

Quashia shook his head. 'I've never seen one like it before.'

I placed it carefully back into the small pruned tree and left it there, watching as it resettled and all but disappeared before my eyes. An Alice-in-Wonderland camouflage.

'What about our own bees?' he nagged again.

'I am considering the idea, Mr Quashia, but, you know, bees and pesticides are not compatible.'

'Carol, bees have been around since forever and folk have been spraying for centuries. You worry too much about nothing. But if you're really concerned, take the hives off the estate in the late spring, drive them high into the mountains and settle them where they can pollinate the wild flowers of the Mercantour National Park, beyond a risk zone, up where François was transferring all his?'

Ah, the ancient practice of transhumance. That was not a bad idea.

'To accomplish that, we would need a licence to install them within the protected park.'

But I felt sure that François, who had placed many of his hives there and always boasted proudly about the fine quality of his wild rhododendron honey, would agree to assist me with the bureaucracy involved in attaining a permit. 'Yes, that's a possibility.'

'I told you, I'll take care of them.'

But what if Quashia should leave for months on end as he had threatened, burying himself deep in the deserts of Algeria, replacing his teeth, or because my farming methods had driven him away? Who would care for the hives then?

Since the earliest of times, the fairs down our way have been a cause for celebration. For a brief spell, the hardships of the earth were set aside, the skin was washed of soil, glad rags were donned and everybody danced, flirted, entertained their children and carried home some hard-earned cash. These rural gatherings were markets, each selling his wares, but they also provided lively social occasions. Even in their twenty-first-century guise, I always thoroughly enjoy them, particularly when I encounter locals conversing in Provençal dialects.

This year, the first of the annual olive fairs, dubbed 'Feast Day of Recently Pressed Olive Oils', a celebration of the latest vintage, was to be held on the second Sunday in March, a little later than usual, in the village closest to our farm. It was billed as a one-day event. I had expected that it would be laid out along the central boulevard that climbed, not quite straight, up from the coast, sliced through the toes of our medieval village and was lined with lofty plane and eucalyptus trees. Its principal square, a veritable dustbowl opposite a dry cleaner's, indifferent baker's and *tabac*, where *boules* tournaments were regularly held, was some way south of the town hall and was used on market days for itinerant commerce and other activities such as the reparation of cane chairs, a service frequently offered by local gypsies. Here was the *place* that every December welcomed our eternally waving Father Christmas. Here, outside the old town, was where the action in the village took place and it struck me as the perfect location for the olive fair. Or, if not here, then why not in the historic heart of the village with its terraced houses, cobblestone streets and extensive views out to sea? However, neither spot had been selected. The appointed locality was a short drive beyond the village's outskirts in an urbanised district where a series of rather ugly apartment blocks had been thrown up. Fortunately for all concerned, it was a glorious early spring day and at half past nine on the designated Sunday morning we set off down the hill, hoping to arrive ahead of the crowds. No such luck! The stalls must have been rigged up soon after sunrise, for a jolly throng was already wandering among them, chatting (in French), socialising, shopping, while the vendors were busily replenishing their displays. The mood was cheerful. We were not there to purchase olive oil or its products. My purpose was to meet with one or two producers, if any should be present, who were farming organically. Before we had strolled more than a few steps, we bumped into friends, Bridget and Luigi. They were brandishing a jar of black *tapénade* and another crammed with large green olives soaked in brine.

'There are several organic stands. See, these!' she cried approaching, knowing for sure what I would be after. 'Further down there on the left and another on the right, and their oils look good, too. Take a look, why don't you?'

'We will, we will,' I answered. 'Have you been here long?'

Like us, they had just arrived but she suggested the festivities had begun somewhere around seven.

'As in earlier times with the descent to market from the hill villages,' I remarked to Michel as we negotiated the length of the exhibitions, through the growing press of people.

'In bygone days, the producers sometimes journeyed all night with their crops on horseback or donkey and cart, descending from the hills, in the hope of earning the cash they needed to feed their families, restock their farms with seeds and replace their worn-out tools.'

My attention was drawn to a stall where a silver-haired lady, accompanied by a middle-aged man with mutton chop whiskers sitting quietly behind her smoking a large-bowled pipe, was selling copies of her own books. We paused to browse. Her husband was her publisher, she informed us. He had given up his career to support her. Her works, *les œuvres*, were local tales and these usually held an interest for me. Extracts had been painstakingly copied out by hand with a nibbed pen on to lengths of paper and pinned to a burgundy velvet curtain engirding three sides of their stall. The language of her literature was French not Provençal. Posted there were sepia photographs of shepherds, stone *côtes*, wind-blustered hamlets high in the middle of nowhere, images from the late nineteenth century up to pre-Second World War. My imagination was fired and she sensed a sale in the offing.

Somewhere behind us a loudspeaker was spluttering out some information that was almost impossible to decipher. A child had been found, was that it? He – or was it a little girl? – could not remember his name, or that of his parents.

'Such a shame!' strolling ladies in summery frocks and cardigans commented.

'If you have lost your offspring' – a crackly description followed of a small boy – 'please come to Marcel's *bar-tabac* alongside the long table where the complimentary drinks are being set up, next to where the dancing is scheduled to take place.'

A rise in the decibel level of excitement. 'Dancing?!' I heard from one or two of the passers-by.

'Drinks, compliments of the mayor, will be served from eleven o'clock onwards. Chilled rosé awaits you.' I glanced at my watch. It was not yet ten thirty, a little early for the first glass of the day, but not down our way. A perceptible shift in the flow of pedestrian traffic took place. About-turns and giggles at the prospect of complimentary wine. I was still with the books and the authoress was growing agitated. Free drinks elsewhere meant loss of sales.

'There are eight titles in the series.' She was shuffling the paperbacks, nudging them determinedly towards me, raising her voice to be heard above the ascending din. 'They are set in the mountains inland of here.'

'What is the subject of the series? Do you want a glass of rosé?' I asked Michel as I fumbled for my purse. He replied that he did not. 'I'll take a copy of each.'

'Provençal life' was the writer's shrill response. The woman took no credit cards, preferred not to be given a cheque and was hoping to be paid *en espèces*. I was a little taken aback to learn that each book was fifty-seven euros.

'Four hundred and fifty-six euros, please.'

'But I don't have that much cash!' In any case, my enthusiasm had dwindled somewhat. 'I might return later' was my feeble exit line, which, understandably, disappointed and disgruntled her.

We made our way in a leisurely fashion towards the table with the rosé because it was also where the dancing was to take place. This promised to be the highlight of the morning. On our way, I picked up two one-kilo bags of dried Provençal herbs selling for five euros apiece, which struck me as a rare bargain. These events offered opportunities of buying direct from the earth though not necessarily at a cheaper price.

'Are you parked up this end?' requested the lady stallholder, still seated and crocheting something white. I nodded that we were.

'Then pay me and I'll store them under the counter, save you carrying them about.'

We agreed to this.

'On your way back,' she added artfully, 'will you bring me one of those free glasses of rosé? It'll be the first time any local official has ever bought me a drink and I'm not turning my nose up at it. I'd go myself but I can't leave the stall. If I were there with the rest, I'd help myself to half a dozen glasses and some snacks. We pay enough taxes, don't you know.'

We promised to do our best. She nodded, wistfully eyeing others wandering to and fro with refreshments clutched between their fingers.

Before reaching the mayor's table groaning with generous portions of local produce, we encountered the first of the organic olive oil vendors. There was quite a crowd gathered about so I could not approach. I waited patiently a moment or two while scribbling down his farm's address. This olive producer was not from our vicinity. I did not even recognise the name of his village.

'He is further west and inland,' said Michel who knew the place.

Inland invariably meant a higher altitude and this, in turn, would possibly mean that the trees were less vulnerable to attacks from Dacus. When I finally managed to shuffle to the front of the queue, the rather handsome young farmer with a shock of dark brown hair combed unevenly to one side, attentive grey eyes and a red kerchief tied around his neck, confirmed that his holding sat at an altitude of seven hundred metres. He was almost at the extremity of the olive line, he smiled, and there were no problems concerning *la mouche d'olive*.

'Mind you, if the summers continue to get hotter, as has been the pattern over the last three years, who knows? The appearance of the fly at our altitude is not entirely out of the question.' He admitted that for all the landworkers, organic or otherwise, climate change and the consequent shifts in plant habits caused by it were matters of profound concern.

'Do you know if there is anyone here who does farm organically at lower altitudes?'

Someone from behind was shoving me impatiently, waving a blue, twenty-euro note. '*Pardonnez-moi, excusez-moi, Madame.*'

I was, rather selfishly, presuming upon this young farmer's precious selling time, and he had travelled far for this opportunity. I apologised. He shrugged his shoulders and shook his head, embarrassed that he could not assist me and that he was putting an abrupt end to our exchange. I stepped aside, allowing the woman with the money in her hand, with her tightly curled hair dyed the colour of golden raisins, to slip into the slot I had been occupying. She was pulling hard on the lead of a silver miniature poodle with a mauve bow tying its fluffy pompom into a vertical pigtail. The bowser was jumping and yapping excitedly at her heels. She gave a great tug and barked at the poor creature who cowered, while I moved smartly out of the way.

Fifty metres further along, we found ourselves snarled up in the crowds milling about the 'mayor's drinks table'. It was actually several tables placed side by side lengthways and covered in paper tablecloths printed with Provençal designs of olives, cicadas and sprigs of lavender. A line of volunteers were doling out plastic glasses of local pink wine while the public queued, in a not very orderly fashion, to stuff their hands from bowls spilling over with salted biscuits, small square cuts of pizzas and, of course, the familiar black Niçoise olives.

Neither Michel nor I was tempted and would have moved on in search of further olive vendors but we found ourselves surrounded, hemmed in, and there was no negotiating a path in any direction. The loudspeaker was blaring once more, to inform those not within the vicinity of the table that the mayor's offerings would soon have been devoured and to hasten if anyone wished to avail themselves of his generous offer. And then the crowd parted just as though Christ were walking on the Sea of Galilee and from who knows where came a file of men and women, all in traditional costume. The pinafored women wore white lace neck shawls with bonnets tied about their chins while some of the men were in felt hats that resembled fedoras. They were wielding arched garlands of flowers and were accompanied by a handful of fluff-faced youths carrying musical instruments. I took the opportunity to dive at the table and secure a drink for the stallholder thirstily awaiting our return. The crowd began to applaud the arrival of the dancers, but when it became clear that no action was about to take place, they very soon returned to the scrum for food. We slid through a break in the wall of people and continued along the street in search of organic inspiration.

There were sweet stalls, one in particular specialising in nougats of every colour and flavour, all pebbled with nuts. Another displayed gigantic copper marmalade-making pans with the sweet dark liquid bubbling and burping within. Cart-wheels of spiced bread were for sale by the kilo, rustic milk chocolate in great slabs to break off and chew, claimed by the '*maison*' to be '*fabrication à l'ancienne*'. There was a young goat-jawed man in a smaller black felt hat sitting alone, deftly knitting dried lavender sprigs into bouquets. He hailed a greeting to a passer-by in Provençal: '*Boun-joun!*'

Provençal, a language of Occitania, once spoken through-out these southern hills and the long belt of coast, the language of troubadours and poets, was outlawed in the sixteenth century (1539) because it was perceived to be a threat to the French-speaking north, the monarchy and French rule. Over the centuries, as French became the lingua franca, Provençal fell into disuse. In the late nineteenth century Nobel laureate Frédéric Mistral and fellow writers worked to have the language reintroduced. Today, it is being taught in certain schools and these local fairs frequently offer an opportunity to hear it spoken.

Robust Italians had driven across the border in a substantial white van and had set up a magnificent cheese stall. They were yelling and calling their wares in yet another language: *Formaggio! Formaggio!*

Alongside the Italians, competition expressed itself in rows of woven baskets packed tight with sticks of handmade sausages. Sausages spliced with mushrooms, sausages flavoured with duck, sausages from Arles, others with Roquefort cheese, even another sampling flavoured with – for heaven's sake! – bilberries.

The *Nice Matin* was stacked, on sale at the *bar-tabac* where a handful of inscrutable old men sat watching the world passing by, sipping their iced Ricard in silence. A baggy-chinned man in blue check shirt and grey hair was being interviewed by a local radio station. Something to do with the electric olive harvesting rakes he was displaying. He clutched a long sprig of *oleae* as he talked, holding it between fingers flailed by land work.

A member of the dancing ensemble, flabby-cheeked in hat and burgundy printed waistcoat, let off an old-fashioned gun that resembled a blunderbuss. He aimed it high into the sky and the explosion resounded and echoed across the blue enamelled morning and caused a few of the crowd to jump out of their skins or giggle skittishly. 'What a scare!' 'Ooh, goodness me!'

I spotted a second organic oil stand and pushed forward. Drawing close, we passed by another olive stall and I recognised the patron by his name, written above in large green letters. This was the brother of Gérard, our miller. 'Look,' I said to Michel, 'how he resembles their robin-breasted father.'

Michel laughed, for this fellow did indeed. He sported a

bristly moustache and black hat with a honeybee brooch pinned to its rim. Was he also an apiarist?

'The two brothers don't speak,' I whispered to my husband. 'In fact, hearsay tells it that no one in the family talks to this sibling. He has been ostracised.' I had no idea why. 'He does not even press his olives at the family-owned mill, imagine that.' I paused in the alley between the stalls to study him carefully. I had an urge to go and say hello, but thought better of it. We had never met before, and I'd had no dealings with him. He was a fair decade older than our miller, with hazel, pedunculate eyes, yellowed and puffed by booze or heavy smoking. An aura of sadness, fatigue, emanated from him. He followed in the olive tradition, the family trade, but soldiered on alone. It was an extraordinary fact that he did not press his olives at the home mill.

Down our way, there existed a profusion of tales of family rivalries where one or another will not speak to a fellow relation. Still, whatever their rancours, they will always continue to buy their produce from whosoever within the family supplies it: bread from the baker if there is a cousin who bakes or meat from one's brother-in-law, the butcher, should that be his trade. According to the Provençal mindset, to purchase outside the immediate family or the hamlet of one's birth is to do business with 'foreigners', even if those foreigners were in fact inhabitants from an adjacent village.

I wondered silently what had caused this man to be so exiled from his kin. He, whose stall displayed olives of every size and variety, stuffed with peppers, garlic, onions, all on sale for sixteen euros a kilo. The oil he was selling was not organic, no more than his brother's was. I also noted that our mill situated high in the hills behind our home was not represented at the feast of the pressing fair today. Did they avoid all possibilities of chance encounters?

The second organic farmer had also descended from on high. His holding was in a vicinity behind Aubagne and sat at six hundred and fifty metres. So, yet again, the olive fly was not a natural pest for him or his trees.

Did he know of anyone at lower levels . . .? As before, the *oléiculteur* simply shook his head. 'Impossible to farm organically if you have to contend with the fly.' And so said the third, the last of the three olive farmers present operating organically. We were out of luck. I felt downhearted, disappointed. Still, we continued on to the far extent of the displays, to a small square where old men with knobbled canes were idling time, gossiping, sitting on benches beneath the shade of the great plane trees, while across the leafy quadrangle were a trio of Arabs, arms on laps, just like the old men, but these North African labourers – two were barefoot with their plastic sandals discarded on the ground – squatted on a bench of their own. Theirs was at the farthest corner from the events. It looked back across to a less than attractive *place des boules*. An unspoken rule, then, that the Maghrebians do not sit with the Provençaux?

A second shot rang out. It was the hour for the dancing. We turned back in that direction, intending to amble towards the car. The plastic glass in my hand had grown warm and I felt sure the liquid was tepid. A small black insect was floating on the wine's surface. As we passed the mayor's table, empty now, its cloths stained by wine splash, I placed the glass back where I had found it.

The dancing had begun. The flautist and drummers were beating out the rhythms while the performers were spinning to and fro, swept along by their own merrymaking. Back and forth they bounced, flushed from exertion and pride, calling out to one another in their local tongue. The crosses worn as necklaces lifted and slapped against plump female flesh as the couples stomped in circles and then a little heel and toe polka or jig, up and down and about the pavement.

Suddenly, I caught sight of the mayor. A big-bellied fellow crowned by a hat with a swish white feather, clad in a heavy, black velvet cloak, he was standing alongside a woman and gentleman. They were also wearing hats with white feathers as well as the gold chains of office, which was a little puzzling. Perhaps they were royalty from another village? Whatever, this trio of civic servants, watching the display, were grouped together in a line, wine glasses in hand. To their left was a 'messenger' in a scarlet satin cloak who was bearing a blue and white banner. All wore expressions of such lachrymose boredom that they might have been attending an execution.

I nudged Michel. 'Carpe diem,' I said, and we burst out laughing. Feeling the sun on our faces, enjoying the happy-go-lucky mood, the exuberant display of the farandole, a traditional folk dance from Nice, we watched on for a few minutes longer and then, while the crowds were clustered together at this one spot, clapping, whooping, calling, we made our way with ease, arm in arm, back to collect our herbs where, fortunately, we found the vendor humming happily, bobbing her head from side to side, sipping one glass of rosé while another awaited her at her side.

'Is it jolly down there?' she asked.

'The mayor's having a whale of a time,' I grinned.

'Oh, he's a miserable old so-and-so that one, face like a battered fish. Probably worrying what the wine has cost him. Anyway, cheers.'

We collected our herbs, bought two jars of honey each weighing in at a kilo, one for Quashia and the other for ourselves, from a beekeeper who lived way over the far side of the Var (or I would have invited him to come into partnership with us), and we said *au revoir* to the fair where the loudspeaker, crackled sound retreating behind us, was announcing that the parents of the lost boy had been located and they had been reunited.

'All's well . . . eh?'

My mission had been a disappointment, but the morning

had been lighthearted and fun. Still, I was pleased to be returning home, clutching our bags of Provençal herbs and honey, alone with Michel to prepare lunch, and consider in which direction we might try next.

The repairs to the garage roof and extension of the terrace were finally to begin. I was up at five to drive Michel to the airport; he was booked on the first flight to Paris. A little after nine, our quartet of Portuguese came belting up the drive in a white, open-backed, beaten-up lorry. Two in the cab, two standing in the rear clutching dozens of upright planks of wood and an army of iron implements that I could not possibly identify. They ascended the hill at a pace that was alarming. It was a barrage rather than an ascent. They were whipping and damaging everything in their wake, breaking and snapping olive branches, newly and lovingly pruned; I heard the rumpus from my den, glanced out of the window, saw the assault and went running down to the driveway, yelling and waving to them as they reached the parking area: 'Stop! For God's sake, slow down.'

Too late, they had juddered to a halt. Broken olive twigs had been trapped between various parts of the lorry's carcass.

'Please, can you take the drive at a more considered pace?'

They did not listen, they could not hear me. They were shouting loudly among themselves, jumping down from the van, unloading wheelbarrows, boxes, throwing the metal rods this way and that. The dogs were fleeing for their lives.

'*Bonjour*, good morning, *bom dia*,' I yelled. 'Please, can you be a little more considerate as you climb the hill? You have damaged the olive trees. Look!'

'Oh, Madame, it's nothing. Just a few twigs' was their infuriating response.

For two hours they hit, slapped, hammered and rammed at the far stable wall, rendering it into piles of rubble. When that first rather raucous movement of the demolition process had been achieved, they jumped back into the lorry and roared off for what proved to be a three-hour lunch break. After their return, I hurried back downstairs to the dust and debris, brandishing one of the broken olive branches I had collected from the tarmac. 'This is not merely a twig,' I emphasised. 'This might very well have represented a litre of olive oil.'

They stared at me with bemused, bloodshot expressions.

'Oh, Madame,' the husky-voiced short one grinned eventually. 'We are Portuguese. We know everything there is to know about olive trees. We grew up among them.' They seemed in a merry frame of mind and no amount of complaining on my part could dent that.

'I realise that you are obliged to deliver all the materials up here and that the drive is steep . . .'

'And the old bus is not what she was. If we don't take it at a lick, she stalls,' one of the others laughed. 'In fact, she's for sale if you are interested.'

'And the planks are tall,' said another. 'They hit the trees, tear at the boughs. *Hélas*, there's nothing to be done.'

'I realise all that. Still, I must ask you please to bear in mind that both sides of the drive are lined with overhanging olive branches and—'

'Madame, we can see them. We all grew up on olive farms.' Now they seemed puzzled by my concern, bemused by my insistence and gazed at me as though I were in a zoo. Feeling awkward, I turned on my heels and left them to it, crossing my fingers that they were intending to show more respect for their work.

'What on earth are they doing to make all that racket?' I moaned to Quashia later when I walked up the hill to give him a cup of tea.

'It's not the Portuguese, it's the new neighbours on the opposite side of the lane. They are preparing the ground for construction.'

'What sort of construction? We live in a green belt. If we put two dustbins one on top of another here we can be fined.'

This was the stone-walled property that sat high on a rocky bluff looking out over the distant sea. A substantial property, this had also changed hands during my absence. We had yet to meet this man who I believed lived alone with two small children.

'And the other folks, they're cutting more trees.'

'Not ours, I hope?'

'No, their own, but they have already taken down seven or eight real old giants. He's chainsaw-happy that biker.'

A refreshing interlude from the construction and concerns back at Appassionata was my visit, alas, without Michel, to a farm that carried both an organic certificate for its holding as well as an AOC for its oil, a rare example.

Since our one-day course managed by the Chambre d'Agriculture back in February, I had been putting out feelers here and there in search of a government-backed body that supported the move to organic farming. The ITAB, Institut Technique de l'Agriculture Biologique, gave me no results when I typed in *oléiculture*, olive cultivation. I searched the site of the FNAB, the National Federation for Organic Farming. They had no solutions for olive farmers. There really were none. Olive farming was a very small slice of the French cake. Even so, I knew that it was an area in which the nation wished to grow, yet nothing seemed to be about to aid and support those of us working in this field who wished to convert to *bio*. And then I came across Agrivert.

Agrivert was a very small organisation, eleven staff members throughout France, whose role it was to research organic alternatives to the more conservative French farming methods. This young team, I had been told, were working in all agricultural sectors except livestock. They were underpaid, understaffed and overworked. I telephoned and was invited to come along the next day to their office, which was in a *boxe*, one of a row of temporary builders' shacks placed within a cavernous makeshift unit, seconds off the motorway near the airport at Nice. One cramped room situated on a landing was reached by iron stairs above a vast open space where, every morning at three o'clock, the largest of all Provençal flower markets opened its doors to every trader, florist and nursery-owner between Menton and Marseille. Flowers, plants, flown in, trucked or shipped in were the point of contact. The horticultural industry was big business along the sea-fringed Côte d'Azur. Over the past few years, clustering around this vast open space, most offices relating to the various local departments of agriculture had been relocated.

I climbed the echoing iron stairs and was greeted by a young lady with Celtic freckles, pale skin and frizzy hair as red as any in Ireland. This was Nadine. She was the technician who had organised this one-day get-together for a handful of *bio* olive farmers. As I moved to greet her, she shot past me, hurtling down the stairs, running out of the door.

'I'll be back' was her parting call. 'Go and join the others. They are having lunch.'

From down below in the ghostly stillness of the empty flower market, she was waving a braceleted arm signalling to an open door. '*A plus tard*, Madame Drinkwater,' she called. See you later. Scrambling in her bag for keys, she dived into a decrepit bronze Peugeot.

I stepped through the open door into the tiny room crushed with seated farmers lunching at an oval table. At the sight of me, they fell silent. Each turned his head and quite literally gawped at me. I felt as though I had walked into a saloon bar in a western.

'*Bonjour,*' I hailed, smiling confidently, feeling daunted, a fish out of water. No one responded. I took the closest empty chair to the door and settled. The men maintained their silence but returned to their food, throwing surreptitious glances my way, sliding bottles and jars from one end of the table to the other as though playing an intense game of poker.

'Perhaps the lady would like something,' offered the youngest who was darting eyes my way with the expression of one who had never seen a woman before.

'Thank you, I have eaten. Please, don't let me interrupt you.'

Nine organic farmers (but only four, I learned later, had succeeded in gaining the official ticket) had gathered together to bemoan their fate, it seemed, and the lousy deals they were obtaining out there in the brutal world of commerce. Their lunch consisted of offerings from their various farms: a cloudy, unfiltered bottle of olive oil, organic brown bread that these men tore at with stubbled, blunted fingers that reminded me of tortoiseshell thimbles, pastes made from their own hand-picked olives and a couple of bottles of red wine that were chasing down the delights of chunks of Camembert and local goats' cheeses. Among them was the very handsome young *stagiaire*, apprentice, who was dissecting a Golden Delicious apple into wafer-thin slices, each of which bore the crescent shape and translucence of a new moon. The men admired his bone-handled knife, which he had bought at a fair in Bordeaux, he boasted, glancing sneakily in my direction.

The accents in the tightly enclosed space were dense, hard to follow, even when they did not switch to Provençal.

I was the only woman in a world of straw-riddled farmers, out at a meeting where women – their kind of women – did not venture, aside from Nadine, 'a technician', and who in any case had disappeared off to the station to collect one of her colleagues arriving from their office in Avignon.

Their silence was eventually broken by a dark-haired, perplexed man, slightly desperate looking, seated opposite me who piped up with, 'All winter I have been asking myself,

Shall I quit now? Shall I give up this lark? I'm barely scratching a living, not sufficient to feed the hens . . .' He spoke fervently, anxiously.

Beside me, with downturned moustache and unruly hair, a goat-chewed individual began to speak, talking fast, tripping his words over everyone else's and, once started, he rattled on, never pausing for breath throughout the remainder of the afternoon. There was no point on any subject about which he did not have an opinion, an answer, a salutary experience to share.

Our Celtic goddess returned with a clean-shaven, dark-haired man in his early thirties wearing a splendidly woven technicoloured jacket and, slung over his left shoulder, a more or less matching textile bag. Here was Vincent. Vincent, who had travelled from Avignon to impart to us his knowledge on the paramountcy of biodiversity in the olive groves, before leading us on a tour of a hinterland farm where, three years earlier, he had planted up bands of more than fifty Mediterranean wild flowers.

'Biodiversity, the variety of life, of organisms, of species, in the habitats within which we are working is a bedrock of our philosophy. If anyone would like seeds from me, I am more than happy to supply them,' he offered.

'Yes, please,' I raised my hand. 'I'll gladly buy some.'

'Oh, they are not for sale. We are giving them away, or we will be in the autumn.'

The fellow alongside me pooh-poohed the idea of planting wild flowers in groves. 'How can you lay down nets if there are flowers everywhere? Impossible! It's not ornamental gardens we are cultivating here, but working fruit farms.'

Vincent ignored the comment and cranked up a very antiquated projector in preparation for a slide show. His soft-spoken live commentary was intended as an introduction, explanation, identification of the various plants endemic to this region, plants that lived happily in olive groves; of the insects, the fauna that were attracted to them, fed off them. 'Honeybees, for example, have a vital role to play in the pollination of myriad wild flowers growing at the feet of the olive trees. And what is lovelier than an olive grove bursting with poppies, daisies, dandelions, sea daffodils, tamarisk, wild almonds, alliums . . . ?'

Unfortunately, every sentence was interrupted by the man at my side.

'I know about this flower. No, I don't think we should be planting that!'

Or 'that one encourages snails and they're a nuisance. I don't agree with you!'

Eventually, the quality of the projection being what it was, along with the bombardment of unwanted opinions, Vincent shut down the equipment and suggested that it was time to set off. Having travelled by train, our botanist lacked transport. Nadine was staying on to clear up and the men were all going on somewhere afterwards. So was I – it was St Patrick's Day and I had accepted an invitation to a party – but I valued the opportunity to talk to this man quietly alone, so I offered him a lift.

We set off into the hills towards St Jeannet and then, before entering the village, turned right, climbing a corkscrew road cut into the mountain's ridged ascent. I asked Vincent how he had come to this work. He was originally from the Seine-et-Marne *département* east of Paris. His father had worked as a manager-labourer on a large conventional wheat farm, intensive farming, heavy chemical use. He had died young from a rather rare cancer. Vincent had gained his master's in agricultural engineering in Paris but had quickly been drawn towards the alternatives. 'I began to look at the figures, the statistics and ask questions. In the United States, more than eleven hundred kilos of chemicals per person are used each year and nearly ten per cent of these are recognised, identified carcinogens. Toxic chemicals in the States alone are causing the deaths from chronic diseases of over three-quarters of a million citizens each year, perhaps more. The region in France where I grew up is one of the heaviest users of pesticides, crop chemicals. I looked at all this and decided that there had to be another way to live. I suppose I am also driven by a memory of my father, ailing, wasting away.'

Towering above us was the mighty Baou of St Jeannet, a wrinkled eminence, an impressive bluff, that roared out of the mountain range. I was trying to concentrate on the road, listen to what Vincent was telling me – I was discovering that the argumentative farmer who had sat beside me at the lunch table, Henri, had been instrumental in Vincent being offered his present job! – and take in the awesome scenery. For no reason in particular, I had never been here before. It was astounding: dramatic fissures split the rock face.

'There are deep caves everywhere here. The inhabitants used to hide in them to protect themselves from their enemies.'

'Do you know this village then?' I asked my companion.

He smiled. 'Yes, it is born of a very ancient settlement, prehistoric. They claim that, even today, there are women living here in St Jeannet who practise witchcraft and they cure ailments with herbs and secret spells.'

I laughed.

'Most of us who respect and use herbs might be so categorised. Warlocks and witches, we all.'

'Do you know anything about a chrysanthemum that is a natural insecticide?'

According to Vincent, the plant was pyrethrum, indeed of the chrysanthemum family. The dried heads of the flowers made a natural insecticide, an insect repellant, but it was not recommended and, as far as he was aware, not available in France.

'Why?'

'It is indiscriminate, kills other insect life. *Pyrèthre* is its French name.'

This news was a blow. We drove silently while I took on board the fact that another avenue had met a dead end. Then Vincent returned our conversation to St Jeannet.

'Once upon a time, all this area was rich agricultural land and the locals took their farming very seriously. Produce was a matter of great pride, until tourism took hold of the coast. But it has a fascinating history. Not far from here, back down the way we climbed, forking left along another cliff road also carved out of the Baou, is a fabulous fairytale castle. If you have time after we have finished, you might want to take a look.'

Because so much of this part of France, the Riviera, the Alpes-Maritimes coastal spread, had been blotted out by over-construction, I was inclined sometimes to forget its distant past. Little of anything that dated further back than the mid-1800s remained.

'And the village has its own water source. It is not connected to the regional water board but fed from springs high within the stark crags of the Baou. It gives the residents a rare independence and there is a quality about the people here that strikes me as unique. I feel they have retained attachments to a way of life that has been sold out elsewhere in these environs.'

We were passing a wooden gate, semi-hidden by tall laurel bushes. This was our destination. I drove on another two hundred metres and found a leafy spot to pull over.

'This couple have eight hundred trees both young and old growing on a soil that is clay and limestone-based. You won't be disappointed,' winked Vincent. Together, we strolled back along the high-altitude lane. I wondered who ever passed this way. A drovers' track originally, indubitably, herding up the Baou to wind-blasted summits.

Perched high on this dramatic mountain, full face to the

distant sea, hidden by neatly laid rows of mature cypress, it was impossible to guess the rural beauty of the property we were about to visit. The house itself seemed modest enough but the grounds were remarkable. Tranquillity washed over me the moment we passed the gate. Vincent and I were the last to arrive. The other men were ahead, walking in a gaggle, moving slowly. Henri was still talking, offering his thoughts to a winding, pebbled lane that led to the house. I shot a glance to my right up the looming precipice. No construction beyond this level, nothing but a harsh climb to scree, boulders, resilient *garrigue* plants, a few holm oaks and pine trees, topped by a blue sky scattered with puffball clouds. Ideal for goats but little else. This was a long way from the capital of the Côte d'Azur, Nice, nestling somewhere at our toes not twenty minutes by car. And it did not surprise me to have learned from Vincent that the folk inhabiting this rock were of an infrangible breed.

A gentle curve in the path brought us to the first of the ancient olive trees. At its side, an *installation* had been constructed out of logs. Head lifted to the heavens, a wooden man was playing a wind instrument while climbing a defunct, wooden electricity pylon. There was something enormously joyful about this animated sculpture.

And then my first sighting of the view: a south-west aspect spread out, a great basin of swooping fields. It was as though someone had unfurled rolls of richly coloured silks that bumped, wrunkled and spilled into the sea. To the left, beyond an extensive vegetable garden where two young women were hoeing rows of lettuces, giggling uncontrollably, lay irregular, sloping patches of vineyards. Vincent had caught up with the men. I hung back – this, the environment, the afternoon, was too exquisite to hurry – descending at my own pace along stone paths that wound through olive groves, both old and young.

It was a paradise, and it brought my heart back home.

'This is returning,' I muttered to myself. Here was peace, a wholesomeness that I feared had been lost to me at Appassionata and I worried might never be retrieved.

Rounding another gentle gradation, I came upon a second *installation*. This one, on a patch of sharp green grass, nailed together out of rusted tins, was a capped fellow wheeling a bicycle, one tin leg bent as though he were about to hop on his transport and be on his way. For no particular reason, he seemed to personify a postman. The artwork was ingenious, a creative celebration of discarded segments of tin and iron.

Lying at the feet of or leaning up against the solid trunks of olives were the triangular wooden ladders that are so representative of Mediterranean orchards. We have one, which I care for proudly though the men tease me about it. I don't care that it is less practical, heavier to lug about (it's true that I am not the one to lift it), and not as weatherproof as its newer metallic alternatives. It represented a picture of rural harmony that I wished to preserve, albeit it in image only.

Nadine had arrived. She was waving, catching me up.

'Un paradis ici, n'est-ce pas?'

I nodded. 'Their situation is high – do you know what altitude?'

She shook her head.

'I doubt that they are troubled by the olive fly.'

'Oh, *si*, *si*, they are. One needs to be above six hundred metres to escape it.'

'Well, then, how are they managing?'

'Have you seen the others, visited the groves where Vincent has planted flowers?'

I shook my head. 'I haven't got there yet.'

'Let's go.'

We picked up pace, also aware that time was limited because the last train to Avignon was at six fifteen, which struck me as incredible but also reminded me that in certain details this southern part of France resolutely maintained its pre-twenty-first-century rhythms. Only tourism requires more traffic. The locals by habit did not travel at night because they were up before dawn, worked their lands throughout the hours of light and went to bed early. Only when descending to markets and fairs did they sometimes travel after night had fallen.

'Over there is where we have installed *Psyttalia*,' Nadine was pointing vaguely to the right.

'*Psyttalia lounsburyi*, the fly from Africa?'

'We assisted INRA, the research centre at Sophia, in establishing him on five farms in this eastern region of the lower Alpes-Maritimes last September. We found the farms for the scientists and we regulate the situation every few weeks.'

'May I take a look?'

She smiled, mildly puzzled by my excitement. 'But there's nothing to see. Even the boxes they arrived in have been taken down.'

'Ah, so it is true, then, that the experiment has flopped?'

'No. Where did you hear that?'

I recounted our outing of weeks earlier. Nadine listened silently and dropped her head, pawing the ground softly with her right boot. 'Is that what they are saying?'

I was wary then, not wishing to become an instrument of perfidy between two bodies both funded by the government. I could see, though, that she was upset.

'When we return back towards the house, we'll make a little detour and I will show you the trees, the branches from which the flies were hung.'

'And where are they now?'

'Underground, we hope. We will only know later, towards summer, when the olive fly appears from its winter hibernation, whether *Psyttalia* has survived the harsher conditions here. It has been a wet winter, unusually so, and that might have been taxing for him, given that he has been reared in protected circumstances and his predecessors were African.'

'Have you by any chance heard of a vegetable and olive producer called Luke?' I gave Nadine his full name. She thought for a moment and then shook her head.

'Curious, I tend to know most of them in this region if they are officially registered as organic. Did you spot the artwork alongside the paths?'

I nodded.

'Ingenious and fun, aren't they? They were created by the manager. He trained as a blacksmith.'

We had reached the lower fields now, where the rest of the group, the farmers, had congregated. I noticed with a smile that the young apprentice was also wearing a striking striped jacket of many colours. All these men were on their haunches studying the ground with gravity. Henri, our talkative colleague, was frowning at the *goutte-à-goutte* pipes that had been wrapped round the small trees.

'They are too tight!' he cried. 'I must talk to Simon. This won't do at all! He will strangle his trees. They will develop deformed. How can he not know this?'

The apprentice was watching me observe Henri. He caught my eye and grinned. 'Where are you from?' he ventured shyly.

'England, Ireland, predominantly Irish, but I have lived over here for a long time now.'

'Is that why you are here, eh? Foreigners like the sun.'

'I married a Frenchman and we bought an olive farm . . .' I fell silent, considering the plight of our olive farm.

The young artisan was from Bordeaux, which explained why his precious knife also carried the same provenance, and he was here studying for two years with Henri. 'I want to be an organic farmer, to help build a cleaner world for our children,' he shared with engaging timidity.

Henri was still banging on about something or other. No one was listening. I felt rather sorry for him. His face was pale, deeply lined and he really did look dog-eared. So much energy, he expended, on knowing best.

The bands of Mediterranean flowers were not so evident in this season. If we had been visiting in May they would have been in flower. Even so, the variety, the diversity, was apparent.

'How do you know which are the bad weeds?' asked the apprentice of Vincent.

'There are no bad weeds. They all serve different functions. Some attract insects, others repel. Each has its role to play in the ecosystem here. Each is of vital importance.'

I lifted my head, listening to gentle wind music in the branches. In the distance was a soft fruit orchard: cherry, peach, apricot, nectarine. The grounds were humming, vibrant with butterflies, wasps, bees. One goldfinch I spotted. Low drystone walls marked grassy terraces, even if some were tumbling and could do with a day or two of Quashia's skills. Clumps of lavender edged the rows of trees. It was an idyll, seemingly of another era, which perhaps was the point. Suddenly, I recalled the groves in Andalucía, regimented armies of trees, not a blade of grass in sight, not a bird or insect working the fields, as well as the soil specialist's warning that desertification was a serious threat to the entire region if the earth and waterways were given no opportunities to regenerate.

Here was its antithesis.

We roamed silently. The brush of my footfall whispered in the grass, weaving my way in and around old masters and the recently imported adolescents. Even Henri had quietened down. We were in orchards cobwebbed with natural treasures; a garden of delight. The Arab or Moorish notion of a garden: a place for solitude, tranquillity, reflection.

And I felt I had much to reflect upon.

How did this proprietor make these groves work, how did he counter the destructions of Dacus? Without a word, we all began our return to the top of the land. Nadine triggered her finger, bidding me follow.

'Let's go this way and I'll show you where *Psyttalia* has been housed; though, be warned, there is absolutely nothing to see.' As we strolled, the house was high above us, beyond an extended terrace where an infinity pool had been constructed. We passed through a darker zone, overhung with tall trees including one or two mighty eucalypts fecund with flower. Here the grounds were timbered and rank, lusty clumps of irises, wild gladioli, *Phlomis*, or Jerusalem sage, rosemary, lavender, thyme. It was mildly damp and giddily aromatic.

The groves in this area were older – this might have been the original olive plantation – and many had fly traps hanging from them, a method used to alert farmers to the arrival of Dacus. Nadine said the proprietors did not run the farm. There was a manager, Simon, born and bred in St Jeannet, whose ethos was pure organic. He had steered his employers in this direction.

He worked with the traps. They alerted him to the arrival of the infestations. 'He follows this with a spray, a natural plant-based product from America.'

I named the product.

'Yes, that's it, though I believe it is being sold under another brand name in France. He also sprays the leaves of the trees, some of the olders, that is, with *argile*.'

'*Argile*, clay?'

'Yes, the product is kaolin-clay based. Mixed with water it is sprayed on the leaves and fruits and creates a barrier that resists insects. It is a repellant rather than an insecticide. It dries to a white powder and leaves the tree a ghostly silhouette of itself. Of course, if it rains the barrier runs off.'

As we passed beneath several of the loftier olives, I spotted veins of white powder on their leaves, reminding me of a clown wearing traces of forgotten make-up. A mask.

'Has this method been proven? Are the long-term effects known?'

Nadine shook her head. 'It is being tested in Italy but nobody is quite sure yet whether detrimental side effects will show up later.'

In my mind's eye I saw a Mediterranean where all olive trees were coated in white. A phantom landscape. I pictured a world where, as protection against viruses as well as the sun's harmful rays, every living creature wore a full protective screen. 'I wonder whether the clay mightn't disturb both the tree's respiratory system and its ability to perspire?'

'Yes, I agree.'

Nadine and I had arrived at the precious spot where the exotic flies had been given their post. As she had warned, there was nothing at all to see. Even the *paniers*, the baskets in which they had been delivered, had been taken down.

'Now what?' I asked.

She shrugged. 'We have to wait till summer to see whether or not they have burrowed below ground or whether they have simply perished.'

'And if they have perished, will that be the end of the *Psyttalia* experiment?'

'No. The researchers will try the same experiments next winter. All is not lost yet. No results, no decisions, for five years.'

I liked this young woman and felt that I had found an ally.

'Will you come aboard?' she asked, possibly reading my thoughts. 'Join the *bio* movement? Hellishly difficult for olive farmers, of course, but we do offer support, advice and encouragement.'

'But this proprietor manages.'

'Yes, both he and his wife are *terriens* with a true respect for the earth but they also have excellent guidance. Simon was suckled on olive buds, I think,' she grinned. 'But the new trap system, the American product, has made a considerable difference here.'

'I understood it killed bees. In fact, is toxic to a number of insects.'

She dropped her head. 'I know, but I guess you have to make choices.'

I considered this, the impact of it. 'Then how can this farm's oil have been registered as organic?'

'The American insecticide is not a chemical. Spinosad is a natural combatant created from organic compounds made by soil microbes.' She glanced at her watch. 'We must go!'

I had recently looked up asbestos and found that it was also not a chemical, but a naturally occurring mineral. How many years had it taken for its detrimental effects to be acknowledged?

I had agreed to run Vincent back to the station. We were behind schedule and I was a little unsure of the route from here. Nadine said he should grab a ride with her rather than risk missing his train. We shook hands and I thanked him for his time.

'How did the Spinosad-based product get its ticket?' I asked him as he prepared to shoot off. He looked away, eyes towards the summit of the mountain, squinting into the light or because he could not face the truth of the answer.

'How has such a product earned an organic certificate? Please, Vincent.'

'It's not a chemical.'

Nadine was calling.

'I have to go. I'll miss the train.'

'Please can I telephone you in Avignon?'

He nodded awkwardly. We said our farewells and I set off myself, pondering all that I had seen and learned. In my distractions, I took a wrong turning and found myself circumnavigating the cliffside in a steep downwards spiral. It seemed to be leading me nowhere and certainly not towards the motorway. I was in a maze of stone. Down narrow and then narrower lanes, erstwhile donkey tracks, chiselled hollows, rabbit runs, sunk between mountain slab, barely the space to swing a cat, until I found myself facing a locked iron gate set within an impenetrable wall of solid stone. The light was fading. Evening was fast descending, darkness closing in around me, and I had no idea where I was. Hemmed in, there was nowhere to make a U-turn and reversing upwards at such an angle was precarious. Uncertain that I was capable of it, I stepped out to gauge the space available and suddenly tilted my head upwards. There, like an apparition beyond the black filigreed gates, soaring skywards, was the castle. Lit from beneath, it was magical and gave the impression that it was floating, drifting off into outer space, rooted nowhere.

My guts were churned up, not least because I was so lost and night was falling. This cul-de-sac was my mind. I knew we would not spray our trees with white clay. It was unattractive, would change the look of the landscape and a tree covered in clay was, to me, an imprisoned being. So, what was the alternative? The American product that harmed the insect life in the groves? I did not want to consider the possibility that the organic olive future being created here in France was based on compromise.

I gazed once more at the castle. According to Vincent, the village of St Jeannet was believed to have been blessed for its fecundity, its bountiful mountain springs. Painters such as Dufy and Poussin had found inspiration here, as well as the poet, Prévert. I would find inspiration, too. There was a route out of this maze; I had only to find it.

I spent most of that evening on the internet – I skipped the St Patrick's Day party, too late home, in any case. On a Canadian website I learned that the recommended organic product was 'highly toxic' to bees exposed to direct treatment, to drift or any residue of the product that might have fallen on plants in blossom. The product's official site suggested that it was harmless to arthropods, while other advice warned that it should not be sprayed near lakes, that this liquid must be kept from rivers and aquatic habitats such as ponds and creeks. It had the potential for run-off, I read.

What does that mean, I asked myself. It meant that the product should not be applied if there is any possibility that it could spill or drip from the trees and settle on the earth, seep into the soil, go underground and into water.

Do not apply if rain is forecast within the next forty-eight hours.

When it rained, the product was washed away. It had been of little use to the fruits and, instead, it would have been washed on to and then into the soil, sinking deep, polluting the water sources that feed wells, springs, rivers.

I laid my glasses on my desk. It was almost one in the morning. Not only was this product recognised as being dangerous for pollinators such as bees, it had the possibility, according to what I had read from several different sources, of poisoning groundwater.

This was not an easy subject. Its ripples travelled far beyond our olive farm and it left me confused about how to proceed, but I believed my resistance was something I should stand up for. That, or leave the farm.

The following morning, a little exhausted, I was downstairs to greet the arrival of our masonry team. Tradesmen down our way tend to commence work somewhere between seven thirty and eight. It was now approaching nine fifteen and there was no sign of them. Of little importance, of course, as long as the work was achieved within a reasonable time frame and was of good order. Otherwise, they could labour away, as far as we were concerned, whenever the hours suited them. I was about to return upstairs when I heard the creaking of our manually opened gates, the spluttering of their truck. At first at breakneck speed, they then slowed at the waving of my arms. I was like someone directing pedestrian traffic, signalling them to take it easy and not distress the trees.

'Good morning, *bom dia,*' from me when they had parked and alighted.

'*Bonjour, Madame,*' the growls of response.

'Would anyone like a coffee?'

Silence.

'José, would you like . . .?'

Four heads turned. Three were José. '*Si, Senhora?*'

'Coffee?'

Each member of the quartet shook his head, which seemed to be a painful exercise.

'Would you prefer tea?'

Again, head shakes.

'Water?'

'No, thank you. We drink the tap water.'

'Actually, I rather advise against that. We are not on the mains system here, and . . . well, it's better to stick to mineral water. I'll bring you a bottle.'

'Tap water is fine, but I'd like a glass of *vin rouge*, please,' called one, he with the smoke-strangled voice.

I stopped in my tracks.

'*Vin rouge?*' It was nine thirty. The sun was still climbing, fingering its way up behind the pine trees on the eastern quarter of the hill.

'It's all he drinks,' grinned another. I had not yet figured out who was related to whom and which of the foursome was not José.

I delivered a tray of glasses and a bottle of mineral water along with one small glass of red wine, which the slightest of the men, whose facial skin was drawn tight like a beaten old drum, downed in one gulp, slapping the glass back on to the tray.

'*Au travail,*' he rasped, while peeling the Cellophane off his Gauloises packet with his teeth.

He lifted a mallet, raised it high above his head, dark, hirsute armpits exposed – he was naked from the waist up –

and began to slam and smite yet another section of stable walls. Powder clouds rose. The building groaned and cracked. This labourer might have been a bantamweight, but the brick-work buckled without argument, tumbling like the walls of Jericho. Back inside the house, in my den, disks were shimmying off the shelves and congregating in untidy streams across the tiled floor. Everything else stacked on the bookshelves – papers, framed photographs, dictionaries, files – was shifting, sliding slowly, subsiding groundwards. The building was almost rocking to the all-powerful bulldozing. Defeated, I closed up my computer. Concentration was impossible. I decided to go to the garden centre instead and then make my way, early, to the airport to collect Michel. If the fabric of the house was not up to this, I did not want to be around to witness the consequences.

At the end of the day, Michel beckoned the team – only three very dusty Josés by this stage – signalling them to install themselves at one of the wooden garden tables. 'Shall we take a moment to outline and agree upon the timescale of the payments and decide what needs to be paid in cash?'

They agreed readily, grabbed cigarette packets and sat themselves down, smiling, nodding, coughing, lighting up.

It was customary down our way to pay 40 per cent of the negotiated fee up front, but Michel usually insisted upon a smaller investment at the outset.

I offered the trio a glass of wine.

'*Non, non merci, Madame.* We are driving; there are police everywhere these days.'

Their words were fact. 'Sarkozy's France,' I lamented.

'*Mais oui, la France de Sarko.* It's a crying shame. You can't get away with anything any more.'

'A chilled soft drink perhaps?'

They shook their heads disgustedly.

Money matters settled, a substantial cash portion handed

over there and then to facilitate the purchase of their materials, also 'on the black' – the Mediterranean *société noire* – they rose, ready to bid us *bonsoir*.

Before they departed, requested Michel, 'An approximate schedule, please. Most of the working materials are here now, is that correct?'

'Oh, *non*, Monsieur, there's the cement mixer, sand, bricks, bags of cement. Plenty more to come and these will be arriving on a daily basis.'

My heart sank.

'Well, the demolition of the far side of the stables has been completed?'

'Yes, and tomorrow we'll begin hauling all that debris away from here. Keep your parking area spick and span,' they grinned.

'*Merci beaucoup*. So, can we assume one month to completion? I think you mentioned such a date to my wife?'

'Well, to be on the safe side, let's say six weeks. Then the entire production will have been put to bed, have no fear.'

'Good, good, that's settled.'

Before they wheeled off down the drive, the pugilist of the party, *petit* José, swung back and said: 'There is no *chef* among us, no boss, you know that, don't you? We are, each one, *le boss*.'

A curious parting gambit, I found it, as was their respect for the drink and drive laws.

The following morning, Michel was returning to Paris for yet another brief spell in the editing suite and I, as always, was up soon after five to drive him to the airport. He would be home before Easter, though, and I was still optimistic that his daughters or one, at least, would come and spend the holiday with us. Michel had a big birthday coming up and I wanted to plan a surprise party, encouraging the girls to be my accomplices. When I returned I found Quashia digging along the border of the cottage garden, planting oleander bushes propagated in our greenhouse, a few lettuces, pots of thyme, rosemary purchased by me the previous weekend. The gate was wide open. Still, I drew up and stepped out of the car. It gladdened me to see him preparing his kitchen garden. I read it as a commitment to his life here.

'Have the Portuguese arrived?'

Quashia shook his head. 'Not yet. A lazy lot, those masons.'

'I hope they have not gone out on the town with our money, or, worse, back to Portugal,' I joshed.

I had no need to feel concerned. By half past nine, the daily mount was underway. In its customary fashion, the gangrenous old banger smoked and belched up the drive, stripping shoots and flowers in its wake, and the men piled out, puffy-eyed, looking exceedingly hungover and unshaven, wired with nervous energy.

Quashia suggested that, to facilitate our olive pruning, he was going to weedwhack the herbage along the terraces. I stayed his hand. The grasses and wild flowers needed to turn to seed first in order that they would regrow the following spring. In any case, they were not excessively long, ankle-length at most. I gave him a brief résumé of my outing to the organic farm, leaving out the use of the questionable product, but he was not convinced.

'But what about the dangers? The fire risks?'

There were precious few fire risks in this season.

'We have everything to gain, Mr Q. We'll cut the land back before summer.'

He stormed off in a huff.

But I took heart from his daily presence and the knowledge that there had been no further announcements since Christmas of his desire to leave for any lengthy period of time, for whatever reason, be it teeth or pilgrimage.

*

Easter. Clarisse, baby Chiara and partner were caught up in the final weeks of the skiing season while Vanessa and her troop were off to Normandy with friends. We had invited no other guests so Michel and I took the opportunity to work on the land. Veils of yellow pollen dust were drifting off the conifers travelling like sandstorms on the horizon, while we laboured at shaping trees. Covered skies but warm, perfect working-out-of-doors weather – housekeeping the land, scrubbing outdoor furniture after the winter, Michel cleaning the pool (still not a squeak from Jacques). A joyous afternoon spring-cleaning. Tendrils of wood smoke fletched and perfumed the mornings. We were pruning, burning olive shoots, the cracked and fallen pine logs, winter debris. I heard the distant crackle of the flames as I clipped concentratedly, far removed on a different pocket of the grounds. When I went to fetch a bottle of mineral water, I spied a newt, a lovely shade of green, padding across the blue mosaics atop the pool. I left it be, wondering if he or she was searching for a site to breed. I spotted several chaffinches, the first in a while, but now they were about, singing sweetly, looping from tree to tree.

After lunch I spoke to Clarisse on the phone. 'Michel's birthday,' I reminded. 'Shall we organise a surprise party?'

She said that '*Papa*' did not want one.

'Has he said so?'

'He told Vanessa when he was babysitting there.'

I was rather deflated by this news. I had been looking forward to the secrecy and intrigue, to the surprised delight on his face.

Several of the climbers I had planted the previous summer, soon after I returned from my western travels, had died off. This winter of heavy rains had been the culprit. Michel was working hard, arms outstretched, lost within the trees' tangle of twisted branches while I trimmed back several of the smaller fellows – it's that eternal reaching upwards that's so exhausting – and then gave it up for the day. Potted and repotted, cleaned up in the greenhouse, planted herbs, aromatics: lavenders, thymes, rosemary. Mud-encrusted my fingers and nails. At the end of one of our long days of work, shattered from pruning – back stiff and arms as immobile as logs – while watering newly sown tomatoes and lettuces in the greenhouse, I spotted a small bird pass gracefully and without hitch right through the centre of a tree I had reshaped that very afternoon. I whooped to myself, delightedly. Yes! I had pruned well!

6

We had taken our time deciding whether or not to fell the survivor of our revered pair of cherry trees. The first had died over a year before. We had missed it greatly for it had been the snowy-white shade in spring, leafy-green parasol of summer and umbrella in winter to our deceased dogs, whose discreet resting place we had created about its feet and where, even as I looked upon the patch now, I could see the settling hump of earth that blanketed our darling Cleo.

I remembered how hard it had been to face the fact that this exquisite tree, not just any nameless tree, had died and needed to be removed; that it, along with the animals it had sheltered, had reached its end. This pair of fruiting cherries was one of the surprises, discoveries, we had come across on the land after we had completed our first major clearance of the hillside. There they were, ligneous pillars of beauty and bountifulness, suddenly revealed in all their glory. Both produced the finest, burgundy-black cherries which, when you bit into them, bled a sticky, wine-tinted juice. In place of the first lost fruiter, I had introduced two small ones. Now, the second of the original venerables was following in his companion's footsteps. Early spring was all about us. The red squirrels were reappearing. The sun was rising higher in the sky each day and everything else had burst into life, but this had remained the same: a dark post of wood with three tall fingers, just as it was when we had cut it back the autumn before. Lifeless, and we all of us agreed that the moment for its felling had finally, inevitably, come.

The men performed the deed and once this formidable fruit-bearer was down and sawn into hefty logs, we found that beneath the coating of bark, which fragmented at our touch, the once sturdy trunk was fretted and alive. Streaming rivers of enormous black ants inhabited networks of tunnels. Ascending from its root system, they had hollowed out the tree's inner timbers and made their nests in a complex display of corridors, passageways, galleries. Entire communities were running frantically, at a loss to understand what had happened to their fine wooden home. I refused to spray them but feared they might take up residence in neighbouring trees – had they been responsible for the demise of these two fruiters? – so we decided to burn the cherry wood immediately, piling up a monumental bonfire. We brought chairs, cups of green tea, dragged some of the slender, withering lengths of striplings and root shoots from the pruned olives and sat together in the late afternoon watching the gradual disappearance of a being that had given us such a spread of beauty in its annual flowerings followed by its exquisite fruits. This was its final burst of light.

Quashia hinted that if I had agreed to spray it, this cremation might have been averted. I ignored his judgements and sat silently listening to its aerated timber crackle and pop. The tree burned all night. I woke once or twice and sat up in bed watching the tiny flames religiously devouring every last splinter. I slipped on a robe and stepped barefoot out to sit by the glow of the fire. Within moments, the two dogs had found me and flopped contentedly at my side.

We had lost Bassett, lost Cleo. The honeybees had gone.

This tree, too, had released its spirit. Surely it was time for us to be moving on as well?

I loved it here more than anywhere I had ever known, but I would rather leave and move inland to another property where I was not faced by a choice that I could not live by. I had confided little of my uncertainties to Michel. He judged my worries extreme. Was he right or was the battle to stay here and live a simple pesticide-free existence impossible?

I watched the flames dancing beneath the clear navy sky potholed with liquid stars. An owl in the highest of the pine trees, overshadowing the cherry now burning at my feet, was hooting. I spotted its partner, flying low with silent beating wings, out on a hunting trip. They had young up there, I was sure of it.

I was recalling all that had brought me here, how I had found myself in this corner of the world fighting for trees and bees and squabbling the odds with an old Berber who wanted nothing more than to earn sufficient funds to send home to his family in Algeria, visit a brothel from time to time in Marseille and otherwise enjoy a quiet life. I had given up a great deal for this existence. I had changed languages, careers, taken on a family who at that time were not related to me, not of my flesh, married a man who sometimes when I looked at him padding about the place in a dressing gown and bare feet I thought to myself: who is this person? What had I and this tall Frenchman, originally from Germany, who remained remarkably attractive even though his hair had gone grey and he was less lean, who worked with unremitting dedication, demanded next to nothing from life except his freedom to make the films that mattered to him, who never or very rarely lost his head as I was prone to do – what had we to do with one another? What had brought us together? How had we found ourselves living under this same flat Mediterranean roof? He, who I had intended would be the father of my children. He talked on the phone in languages I did not understand, recounted tales to me in our intimate moments of a childhood that was far removed from my own. Our parents had been on opposite sides in a war! What unexpected turn of fate had delivered us here together, living this life to which neither of us was born, but which had given us such immeasurable satisfaction? And what would happen now if we could not see eye to eye on whether we should stay or go? How much of Appassionata was woven into the fabric of our love?

When we first found Appassionata, I had been looking for a new way to live. I had been searching for many years for my House by the Sea. I wanted to be able to chill out in between acting roles, escape city stress and commitments. It had not been clear to me then that I was hungering for a life in harmony with the world around me, the environment. Appassionata had seduced me, had seduced us. It became the centre of my world instead of a bolt hole on the periphery. I fell in love with the Mediterranean, with its nature and seascape and perhaps above all with its olive trees. The olive tree and all that it represented; the history, mysteries it carried within its DNA. I had never intended for all this to happen. I never intended for this *modus vivendi* to take such a hold, for it to kidnap me from what I had always perceived as my *real* life, but at some loosely evolving point, I cannot precisely determine when, I took my first steps along a new path and I kept walking. It seemed to be the 'right' route. At every crossroads, when opportunities to turn left or right, to veer off from this path, were offered to me, I kept going, immersing myself ever more profoundly in the universe of the Mediterranean and its natural surroundings. I grew more conscious of my environment. Nature became my companion, particularly when it became clear that I would bear no children, and I was happy, content in this life. At one with it. But now, I found myself at a crossroads that left me uncertain, troubled. To go forward, which was to say to stay and fight this environmental battle, demanded tools, expertise, methods, a system with which to combat the olive fly, but nothing suitable was on offer. On the other hand, to stay and ignore the choices to be made . . . to remain in the system, keep our AOC, keep faith with the local agrarian community and their traditional methods, enjoy our splendid olive oil, yet all the while knowing that the spray that fell on to these acres was endangering Appassionata's ecosystem and would eventually seep into groundwater and contribute, even if only to a minuscule degree, to the degeneration of the earth, to the disappearance of honeybees, endangering other fauna and man . . . I felt unable to accept this.

So, this home, this idyll, this patch of paradise was calling upon me to champion it. Who would know if I simply turned a blind eye to this dilemma? Michel and Quashia, yes, but Michel was not as extreme in his philosophies as I am apt to be and Quashia was dead set against my 'nonsensical reasoning'. No one else. Only me.

I must have nodded off by the fire, gazing into the remaining cherry embers, because I was woken by the vigorous barking of dogs. I looked about me, whispered their names. They had decamped while I was dozing. The fire had died. I shivered, felt chilly, sat up to listen. I assumed the alarm had been triggered by one of the neighbours' cats. They used our land as nocturnal hunting territory. In broad daylight, due to the dogs' bounding presence, they did not approach, but at night they were bolder and stalked their prey even to the perimeters of the house. It was too late for the boars. They were evening hunters in our neck of the woods, not dawn raiders. A cat or rabbit, then, had disturbed the dogs.

I stood up, stiff from having been curled into the chair. Everything was quiet again, still. I trekked back towards the house, the open bedroom doors, hoping to grab some sleep but then a yelping pierced the darkness, high-pitched, rather like a scream. It was not a screech owl, I felt sure, but an animal in pain. Once my ears were attuned, I realised that there were several creatures. I recognised both Lola's and Homer's barkings but there was another, unfamiliar to me. I slipped barefoot on to the upper terrace; the tiles were warm beneath my feet. I called softly into the night, beckoning the dogs. Creatures answered me, feeding owls perhaps, but not our two hounds. Neither came. Occasionally, they got loose and slipped out beneath the wire fencing where the boars had chewed holes to create access. I went for sandals, in search of them. Up and down the drive in the darkness I pursued them, but the barking had gone quiet. All had gone quiet, as though the night were studying me with watchful eyes. I sniffed, listened, waited. Perhaps it had been a fox? But where were our faithfuls? I searched everywhere, back and forth, circling. No sign of them. I wasn't adequately dressed and did not feel inclined to penetrate the extremities of the property, mostly for fear of encountering nocturnal creatures I was unfamiliar with and because I had foolishly forgotten a torch. Eventually, I returned to the house. They would slope home when they were ready, as they always did. The other barking I had heard might well have been a fox or the Rottweiler across the lane. If our two had got loose they would have enjoyed positioning themselves outside his gate and tormenting him, harrying the great brute and for this, for sure, I would be in serious trouble with the neighbour again.

I had overslept. Quashia was calling to me. It was gone eight. I padded downstairs to make coffee, to serve up the dogs' breakfast bowls.

'Take a look at this.'

'Oh, my God!'

Homer, our German Alsatian, born on the farm, and the light of our lives, had been gruesomely attacked. His face was gashed and lacerated. Had he been clawed or might it be teeth marks? Shivering with shock, he had secreted himself at the foot of one of the cypresses, cowering in a corner. His wounds were now packed with congealing blood and soil.

'He's been mauled by a wild boar.'

But were the boars to blame? It must have been Homer I had heard yelping, then.

'I'll call the vet.'

Fortunately for us, one of the two new vets lived only lanes away – she and her partner had purchased the little riding stable on the far side of the hill – and she agreed to drop in on her way to the clinic. The incisions, she diagnosed, had been made by teeth.

'I would say he has been in a fight, possibly, almost certainly with a fellow dog, a large one, but it is most curious that there is not a scratch on him elsewhere. Poor fellow is feeling very sorry for himself.'

Homer was given a tetanus injection, a five-day antibiotic course, oodles of tender loving care and extra food if he wanted it, but this he refused. He ate nothing for several days. He was sulking, in pain. He was discomforted by the bandage wrapped around his muzzle. As soon as I had replaced it and turned my back, he tore it off and dragged it through the dirt. The filthier it was, the happier he was. His pride and swaggering young manliness had taken a real blow. I did not know whether to laugh at him or cry with him, but I was determined to find the culprit who had inflicted such a wound. It was time to pay the neighbours another visit.

Armed to confront the worst, I strode down the drive where, to my utter amazement, I found himself beyond their gate, in body-hugging black leather suit à la Johnny Hallyday, splayed out on the ground. His bike had fallen on top of him. He reminded me of a beetle turned upside down with legs wriggling in the air attempting to right itself. The difference was that this fellow was face down and could not extricate himself.

The sight of him took me aback.

'Can I help? Are you all right?' Clearly not. I drew closer, gingerly. 'Shall I ring the bell, call your . . . wife?' Was she his wife? I had no idea. 'Do you want a hand?'

Slowly, he turned his head, attempting to focus, to identify this woman looming over him. It was not yet 10 a.m. and he was smashed out of his mind. Whether he had attempted to go out and had fallen while mounting the bike or had been returning after an excessively heavy night, I had no idea. Whichever, he was incapable.

I dithered idiotically, not knowing what to do, forgetting about the plight of poor Homer. 'I'm going to ring the bell,' I announced. I took a step. The Rottweiler cantered up to the fence and shoved his snout through a triangular gap in the wire. He was barking, snarling. I realised then what had happened. Homer must have stuck his nose through the fence and the monster had gone for him.

'Leave me be,' slurred the neighbour, aggressively. 'Can't you sheee I'm fine? Fuck off.'

'Right, I will.' I turned on my heel. I had plenty of concerns of my own. If, as I felt sure, this Rottweiler had attacked Homer, I would make sure the dogs never encountered one another again.

As I plodded miserably up our hill, the white lorry belonging to the Portuguese came grinding up behind me. The men were whooping, laughing like schoolboys. You might have thought they were off on holiday. I stepped to one side and let them pass, watching as yet another branch cracked and snapped from the fig tree. Quashia had tied the hanging boughs of the olives back from the driveway towards their central trunks; not an ideal position for their development, but the alternative was far less appealing.

'Good morning,' I waved as I puffed up to the parking and the men, three only, were unloading tall iron rods. I was not in the mood for noise. I called to Homer who trotted slowly after me, bandage unfurled and hanging, and I started towards the house.

'Madame, Francisco, he cannot see,' one of the Josés announced.

They had lost the fourth member of the quartet, due to an eye infection, he came to tell me.

Blind drunk, I giggled to myself under my breath. 'Oh, dear, I'm sorry to hear that.'

'Today, we need you to clear the garage so that we can knock out the remaining ceiling.'

'Now?'

I called to Quashia for assistance. What a chaos of moulderingness. How had we ever accumulated so much junk? When we had finally acquired Appassionata and I entered this garage, I remembered how fascinated I had become by the piles of detritus, forgotten lives relinquished there, and now, here we were, little better. I tucked Homer into a corner on a blanket and went to work. Lack of sleep was kicking in. Decision-making was not at its keenest.

'Shall we sling this?'

I was thumbing through the pages of a film brochure, glossily encased pitch material for a project that we never raised the funds to shoot, but it remained a terrific idea. For a second, it made me wistful. Dreams that had hit the dust. 'It's hard to know where to begin.'

Quashia shoved me out of his way. 'I prefer to do this myself. If I wait for you,' he moaned, 'the way you dither and moon over everything, we will never get through it all. And those blasted men'll be here all year. They are destroying my trees, they show no respect. And they drink!'

'In five weeks, Mr Q, they will be gone and you will have a fine garage with a dry ceiling.'

As I stepped over plaster lying on the floor, I was stunned by this room of lost things with its discarded souvenirs of earlier selves decaying in the damp. I rose up from my haunches and slapped Quashia on the back. 'You're right; you are far better placed than I am to clear this lot out. Shall I leave Homer with you or take him?'

'He's fine with me.'

I passed the remainder of the morning in the *hangar*, shaded by fruit trees, preparing for the stockpiling of the clutter Quashia was not throwing out. When I stepped out into the morning air to stretch, I noticed that the delicate green leaves on the peach, apricot and nectarines were looking as though they had been embroidered, stippled with a rather attractive red along their upper surfaces, and then, cinders were falling around me. Smoke was rising. Where was the smoke coming from?

And what was the rumpus?

The trio of Josés were setting up a table at the edge of the parking area within the long shade of the cypresses. On it stood four bottles of wine: red and rosé. Three upturned dustbins, ours, with a plank of wood, served as a bench. A small grill, filched from one of our defunct barbecues, had been placed on the flames of a fire burning in the grass. Sausages, steaks, cutlets were sizzling and spitting. Lola, our rotund mother Alsatian, was slavering at the men's feet.

Petit José, who must have been handsome once upon a time with eyes as cornflower blue as a girl's, but now marbled by hard living, cut himself a hefty chunk of bread and then another for the greedy dog who gobbled it whole.

'Please don't feed the dogs!' I called.

He nodded, not listening, now pulling corks, splashing deep red Bordeaux into generous-sized plastic tumblers. Behind them, the appetising smoulder of meat griddling on an open fire.

I returned to the house to fix myself a snack and there, on my computer, I found a most unexpected communication. An email from America, from the head office of one of the six major agrochemical companies; they who produced the 'organic' answer to our pesky fly. The sender said he had heard that I was looking for a non-chemical treatment for our groves and he would like to help me achieve this goal.

We want to do everything within our power, Carol, to make sure you get a bumper crop this season.

I was stunned. I read it over several times and filed it. A response required consideration. How had the sender heard of my quest?

I observed the post-prandial pleasures of cigarettes and coffee from the terrace above but said nothing. My mind was on the American note. *Petit* José had finished washing up their plastic plates, cutlery, frying pan, in a bucket Quashia used for transporting cement. Their wine glasses remained on the table, three-quarters supped in the shade. No doubt they would be returning to them throughout the course of the afternoon. Now they were settling to a smoke while their coffee was heating on the open fire that had burned a black circle in the grass. Quashia was still hauling and trawling in the garage while I, overtired, intended to complete some desk work. Lola and our convalescent, Homer, had been seduced. They lay at the feet of the Latinos, hoping for titbits, refusing to respond to my bidding.

One of the men, plump, black-haired 'big José', was not preparing for a nap. He slugged back his cup of caffeine and set off in their lorry to discharge some of the mounting piles of rubble. Only a few days earlier they had been so conscious of the driving laws yet today was another day. Did they think the police were also taking siestas?

But the work continued, or it would once the vino had been slept off.

Quashia was yelling from the garage, yelling unnecessarily loudly. To disturb the snoring men, perhaps?

Down I went.

'Do you want to keep these gardening shoes – they're worn out – or should I just chuck them?'

'Oh, bin them.' As I turned to go, he grabbed me by the arm and dragged me deeper into the smelly recesses of this stockroom.

'Six!' he hissed from within crepuscular gloom and rancid scents.

I feigned ignorance.

'Three of them have downed *six* bottles of wine.'

'Four, but they replaced two into their lunch satchels,' I lied. Our man was not fooled. He shook his head, jettisoning a torn box into bulging dustbins.

'They've purloined three dustbins and they're incapable of work.'

'Have faith, Mr Q.' I returned to my paperwork.

Forty-five minutes later, a rap on the door. Slender, topless *petit* José had a face on him that signalled trouble. His breath could have been set alight by rubbing two sticks together.

'We have fallen upon a problem,' he rasped solemnly.

Bidden to the site, I found only one other José scratching his head and staring at a tape measure.

'What's the problem?'

He looked up at me with eyes that were glazed, bloodshot. I threw a surreptitious glance towards their shady lunch table where all glasses stood empty. Was it conceivable they had consumed six bottles?

'The walls are different sizes.'

'That's impossible!'

The little fellow (the only one who did any of the heavy manual work), now at the rear of the two-stable space, was clambering over materials, pacing from one side of the roofless expanse to the other. Arms outstretched, he reminded me of a small boy playing aeroplanes. 'Here, Madame, it measures five metres ten while, at the front, it is four metres eighty-six. There is a difference of twenty-five centimetres.'

'But that's considerable,' I cried.

They nodded gravely. 'Nothing for it but to tear down this wall as well and rebuild it.'

'But the wiring, piping, electricity, water, all is within it,' I contested. 'And you were quite specific at the outset that it could stay. I'd need to call in both the electrician and the plumber.'

'Then you must.'

Michel had warned me, as he stepped out of the car at the airport, that we would be bound to encounter a minor hiccup or two and that I was not to concern myself. 'They can always be resolved.' But this struck me as more than a hiccup.

I tried to find Quashia to see whether he had any thoughts on the subject, while the three men – the third José had returned with lorry from the builders' dump – were squabbling in their native tongue, roaring and huffing like mad bulls, snatching the tape measure from one another like children fighting over a toy.

'We're moving! I can't stand this any longer,' I muttered to myself.

It seemed that the work would involve more than simply tearing down the sole remaining exterior stable wall. It would also involve bringing in an earth digger, shifting tons of soil, drilling into mountainside limestone and dismantling the stone steps that led us to the upper level of that east wing, to the terrace behind our bedroom, which was, within the terms of their agreement, to be cement-filled and then tiled. This was to be the exterior of the flat garage roof. The 'crooked' wall was intended as the structural wall.

'Beaucoup d'argent,' they were threatening, shaking their heads.

But of course.

Had this ploy been up their sleeves all along, once they had agreed to lower their price by so many thousands? Was the wall really as wonky as they claimed or were they too drunk to measure accurately? I had no idea.

Quashia had disappeared, no longer hauling garage innards on to the driveway. Every article that was not to be thrown away was to be rehoused wherever a covered corner could be appropriated. I doubted we'd ever find anything again. I was standing in the midst of chaos with men yelling at me to make decisions. Eventually, I located Q over near the ruin, planting an avocado tree that had been ailing in the greenhouse.

'How are you getting on?' I called as I approached, wondering why he had abandoned the clearance chores. 'I need your advice.'

He shook his head. His saggy face looked tired and old, his eyes faded by age.

'Everything fine?'

He shrugged, ignored me; displeased. We discussed the needs of the tropical plant for a few moments, remaining on non-confrontational territory. Its lower foliage had yellowed, browning at the tips. I had been concerned that it was pot-bound and would die.

'It already looks healthier,' I encouraged, 'being out in the fresh air.'

Our chosen spot for it offered plenty of growth space with nothing but a terrace of vigorously blossoming apple trees for company one level lower.

I urged Q to return to the work site and give me his opinion on the halted proceedings. Crossing from the Second Plot towards the east side, the grass and wild flowers more than ankle-deep now, he said to me solemnly, 'They are not like us, those men. They drink. Have you seen the bottles under the tree?'

I made no response, aware that this was an area where we did not see eye to eye. Drinking alcohol offended our loyal friend's Islamic sensibilities so we always preferred to be discreet about it.

'Those men are troublemakers, not masons. I'm a mason. The job needs concentration, dedication. You should have let me do the restoration.'

Had Quashia felt that by bringing in these professionals we had undermined him?

'You have so many other responsibilities, Larbi.' (Rare for me to call him by his first name.) 'You work too hard as it is. And we don't possess all the machinery this job requires. We would have been obliged to hire it and that would have made it prohibitively expensive.'

'True. I lack the tools.'

This seemed to satisfy him, for the moment at least.

By the time we had returned to the site, the calculations had been discussed and the cost of rectifications assessed. Quashia took one look and testified against dismantling the offending wall. 'This property was constructed at the beginning of the twentieth century. No one expects it to be perfectly straight.'

The masons, jumping up and down, hotly rejected this argument.

I stood back taking in the scene, all too aware that the differences of opinion being voiced were hiding underlying prejudices. Religious, cultural differences. These hell-raising Portuguese judged our loyal Berber an orthodox Muslim and a peasant who was ill equipped to pontificate on the finer points of their art, while he judged them lazy, decadent. Over the coming weeks – months at their current speed – I would be obliged to find ways to keep the peace.

'This wall will carry the weight of all the work from here on in. It must be reconstructed.'

'I built every one of my children a house in Algeria. I've never heard such nonsense.' Quashia never raised his voice, but he was tilting in the balance and stomped off in disgust. It was clear that he believed he could have achieved this work alone if we had purchased the equipment. Michel and I knew he could not.

I returned my attention to the Portuguese and asked outright what the cost was going to be. The difference would be an additional five thousand euros for dismantling the wall plus the removal of earth, digging machine, reconstruction etc., etc.

'This needs my husband's approval,' I told them. 'Are you able to continue until I have an answer?'

They assured me that they could and I agreed to discuss the matter with Michel who would be home the following evening.

Alone in my office later that evening, I opened up the American letter, perused it again and replied: *I wonder how you know of me? I would be interested to hear what you suggest.*

I read it over, sipped the tea on the table at my side and pressed SEND.

March was nearing its end and with its departure came more rain. Q and I were relocating essentials to the *hangar*. Outside, the young, damp leaves on the soft fruit trees, still mottled by burnished markings, were now curling and deformed as though crippled with arthritis. This was leaf curl, I learned. *Taphrina deformans*: a fungus that occurs in spring. The rainy days that we had experienced during the early development of the foliage favour this fungus, which can eventually stunt the growth of the tree and radically reduce fruit production. I had been travelling the previous year at this time so did not know whether it had already taken hold or whether we were seeing it on our farm for the first time. Quashia claimed that this was its second appearance. I heard the edge of accusations returning.

A copper-based fungicide was recommended by our nursery.

'Any organic suggestions?'

Their answer was negative. I decided to let it be.

'If we lose the fruits, so be it.'

Quashia had grown two of the peaches and one apricot from stones and was particularly proud of them.

'But the fungicide would not be applied until after the leaves have fallen in the autumn, Monsieur Q, so, please, let's not argue about this now. We'll find an alternative solution and we'll cure this.'

'It's like everything else, Carol. You are simply not practical. You cannot save a leaf or organise beehives. I see no point in remaining here. My work is wasted.'

I sighed, feeling defeated, dejected.

Our man absented himself for a long weekend and the Portuguese were unable to work. Thursday had brought in the spate of bad weather – no more cement-mixing machines, yelling, hammering, until, at the soonest, Monday, when the barometer promised clearer April days. Michel was down in Cannes participating in the annual spring television festival, which had fallen early this year, and I was alone with the dogs – Homer's face had all but healed – sitting by a log fire, in candlelight, with a glass of white wine at my side, enjoying the stillness, the calm, listening to logs crackle, the muezzin beckoning the Arabs across the valley to prayer.

The world was honouring Earth Hour this evening and I had signed the farm up for it. Commencing at eight thirty for one hour on this relentlessly wet Saturday, the lights would be switched off. I had been anticipating spending these designated sixty minutes out on the terrace, stargazing, tuning into the planet, but the non-stop rain had driven me indoors. Before Michel headed off after breakfast, I had reminded him that I and the farm would be in darkness for a spell. He was booked for a business dinner and would not be back until late, arriving with a house guest, a gentle Chinese colleague and client of his whose name always confused me: Chowguong (or was it Guongchow?).

Earlier in the afternoon I had run through the rain down

to Quashia's cottage and disconnected the time switch; I had reset our automatic clock that floodlit the terraces around the house; I had fed the dogs, brought them in with me, switched off the music, closed down my computer, photocopy and fax machines and lit candles. The fire was roaring. I refilled my glass, reminding myself that although the fridges were still operating, they would need to stay closed throughout this hour because their interior bulbs came on when their doors were opened. I had thought of everything.

I settled myself on the old spreading sofa alongside the fire and began to sink into my thoughts when, suddenly, the exterior lamps all around the property alighted. I must have miscalculated the programme. I hurried downstairs to the garage to reset it, but it refused to disengage. Sufficient wattage to illuminate a miniature Eiffel Tower was blazing out towards the sea from this eco-friendly property. I twisted and turned the dial back and forth, breaking two nails, until, eventually, I slapped off the trip switch at the mains and cut off the electricity supply to the lower storey of the property. Garage, pool, drinks fridge, exterior lamps and, still weathering it out in the stable block, the washing machine. Every appliance, every bulb on the lower storey had been disengaged.

I sloshed through the rivers of fallen water, showing no signs of easing, and reinstated myself on the sofa. The logs were crackling and across the valley the muezzin remained in full voice. Still and peaceful was the world and I was grateful for the calm.

Eight thirty.

I closed my eyes; my head sank back against the cushions in meditative mood.

Suddenly, my mobile began to ring and I almost fell over the coffee table in the darkness trying to reach it.

'We're here! I'm down at the gate.' It was Michel.

'What time is it?'

'Dead on nine, we're early. See you, love you.'

'Michel! Michel—' The line had gone dead and I could hear the chug of the old farm bus making its way up the drive. I slipped on shoes, out into the soaking darkness, down the stairs, to remind him of the reason for the blackout. Both men piled out of the car.

'Hello,' I waved to Guongchow – no, Chowguong, who does not speak many words of any language other than Cantonese.

'Welcome, welcome, nice to see you again.' Huddled in a coat, he hurried from the car wheeling a massive suitcase and stood before me bowing.

'Sorry about the lights.'

Michel was now at my shoulder. 'Has there been a power cut?'

'Earth Hour, don't you remember? I mentioned it this morning. Don't let's stand here. It's pouring.' I was attempting to direct our guest upstairs rather than to his bedroom at pool level because I had not left any candles in his room. I had not been expecting them back so soon, not till after the lights were back on. 'Let me take your case, Guongchow. Go with Michel and get warm by the fire.' I was speaking slowly, hoping to make myself understood.

'It's Chowguong, Carol,' Michel growled at me.

'Yes, sorry, Chowguong.'

Chowguong was bowing, smiling, attempting not to express confusion, still clutching his suitcase, which I was trying to wrest from him.

'What has happened to the electricity?' insisted Michel.

'It's Earth Hour!'

'Oh, God, that's right,' he sighed. After a tough day of business he had, doubtless, been hoping for a more convivial welcome.

'Guongchow, sorry, Chowguong, go with Michel, make yourself comfortable. I'll put your case in the room and fetch some candles.' I was communicating slowly and with sign language. He was still bowing, smiling, smiling, no notion of what I was talking about. Now, he made the gesture of hands to the side of his head as one would tell a child that it was bedtime. 'Sweeping, pliss,' he said.

'Carol, he's jetlagged and wants sleep. Can't we turn just one light on?'

'I'll get candles.' I shot upstairs by the sodden, waxy-leafed magnolia tree, grabbed boxes of candles, matches, and was about to redescend when the men entered transporting heavy loads of damp shopping for a cocktail party we were giving the following evening.

'Ah, good, you're here.' We stood in the hall a little awkwardly. 'Come in.'

Chowguong bowed again. 'Sweeping, pliss.'

'Follow me. Look, I've got candles! In twenty minutes, lights! It's Earth Hour, Guongchow.'

'Chowguong, Carol, why can you never get his name right?'

'Sorry. Earth Hour, Chowguong, in China, too. All over the world.' I was describing with my hands a big round globe. 'Lights out – gesture for an energy-conscious planet.' I was standing in the darkness, water dribbling down my face, a Give-Us-a-Clue pantomime actor, Chowguong, my bemused audience, nodding and bending. 'Yes, yes, see me.'

'Thank you for understanding.' I escorted him into his room and lit candles by his bed. He waited at my side, still nodding. 'Henkyou henkyou. Sweeping now.'

'Sleep well.' I pointed to my watch. 'Fifteen minutes and lights on.'

'Henkyou.'

Upstairs, I shut the door, closing out the weather, shivering. Michel was in the kitchen, about to open the fridge. 'I need a drink.'

'Have red, or share my white!'

'Have we run out of rosé?'

'No, but we . . . Please, don't open the fridge . . .'

'Why not?!' he begged, exasperated.

'The bulb ignites wh—'

'I don't believe it!'

'Please, it's only twelve more minutes.' My husband ran a hand through his moist hair. Suddenly, his mobile began to ring.

'Yes?'

It was Chowguong ringing from his Chinese mobile. An international call from the room directly beneath us!

'Is he all right?' I hurried downstairs. All the lights were ablaze on the lower floor. I stood gaping, horrified, as our Chinese guest was out in the rain in complimentary white hotel slippers I had left for him, jumping gleefully between the terrace pillars. 'Lice working now. Lice working. Chowguong fix lice.'

He must have read the distress or some other such reaction on my face because he fell silent and flipflopped towards me, slippers muddied, soaking, anything but white. 'No good me, Chowguong?'

'It's fine,' I smiled, trying not to laugh.

'Prefer you lice problem?'

'It was Earth Hour, but . . . hey . . .'

'Ooooh, noooo, now, see me. Earth. Lice. Lice too many . . . sorry, me sorry, sorry, sorry. Lice not good. Me sweeping. Better. Sorry.' And he bowed yet another goodnight and disappeared to get a good night's 'sweep'.

'Goodnight, Chowguong,' I called after him. 'It's all fine.'

Upstairs, Michel glowered.

'It's nine thirty. Let's open the fridge and I'll pour you a drink.'

'You know, *chérie*, you are worse than a fundamentalist sometimes.'

*

A longing for a property sufficiently capacious to accommodate Michel's family; a home where grandchildren would visit and where I did not have to face the complexities of olive farming continued to offer itself as the only real solution and so I decided to begin to search. My standards were higher than in the days before we had found this scruffy old farm because I was all too aware of what I was going to lose as well as gain. Few properties ignited a flicker of interest on my part until the file came through by email for a 'chateau' close to Avignon. Estate agent-speak peeled away, it was in fact a three-storey *bastide* with twelve bedrooms situated within three and a half hectares of forest, grasslands and fields with aromatic garden, large natural pond and swimming pool, so claimed the brochure.

Perfect.

'Does it have olive trees?' I asked Thierry, the Swiss-French broker. The agency handling the file was based in Geneva.

'No, I am sorry, *désolé*, no olive trees. No crops are farmed on the estate at all though the region does carry an AOC for olive oil should you wish to plant some.'

It had been a working farm until the Second World War, but its vast acreage had since been divided up and sold on, a history not dissimilar to Appassionata's.

'It carries no agricultural ticket with it at all so, unfortunately, the annual taxes are assessed accordingly. It is used as a holiday home for the family of one of France's noblest champagne dynasties.'

No further responsibilities. I could grow flowers and vegetables, entertain guests and surround myself with grandchildren by the busload.

'There is, however, a vineyard abutting the property that is also for sale. The *viticulteurs* are happy to either dispose of it to the incoming proprietors of the chateau or they'll keep hold of it. I don't know if that would be of interest to you.'

I preferred to wait and see, was my response.

'One small snag,' continued Thierry. 'I should warn you. There is some unsettled business about a troublesome neighbour. I don't know the details of the story myself, but perhaps you might want to make an offer for that property as well.'

Troublesome neighbours. I was not so sure about that.

I set off alone for the Montmirail region on a warm weekday morning. I had arranged a rendezvous point with a local agent, an intermediary, at the motorway exit for Avignon North. From there, I would follow him to the estate. I had seen a few photos on the internet and I was holding out high hopes for this property. My optimism about Michel's reaction to this act of mine was not so buoyant. I had proposed we find somewhere that did not present us with our current dilemmas, but he had refused to engage in all such conversations, concluding that he had no intention of visiting other properties. Hence my solitary outing. In any case, he had returned to Paris to direct the sound recording of one of his films and would be away for the rest of the week.

I arrived within fifteen minutes of my prearranged eleven thirty slot, in spite of heavy traffic and roadworks on the *autoroute*, and shook hands with a short, wiry Spaniard in gold-rimmed Cartier glasses. We set forth into the countryside, having agreed that I would follow his silver Audi. I had never visited these foothills of the Mont Ventoux before.

'It'll be flat for the first quarter of an hour and then a land of wines and lacy mountains,' he had promised.

At first, it was not merely flat but ugly, overrun with low-density urban sprawl that, until a decade or two earlier, would have belonged to Nature, to the creatures of the fields. Crossing from one A road to the next, we turned left off a roundabout and proceeded along a narrow B road flanked by meadows. From here on it grew lovely, winding lanes with early season vineyards, abutted by inviting tracks that led to undisclosed destinations. Beyond the windscreen, clouds were moving quickly. I longed to stop, to investigate those donkey paths. Once or twice I slowed, drawn to peaceful plains, hankering to be on a bicycle, before I realised I was losing my guide.

Plane trees lined the avenues, patchy solid trunks, overhanging boughs lacking leaf. The seasons came later here, inland. Still, that eternal image of endless Provence was evoked, bathed in summer and dappled shade.

Yes, I like it here, I said to myself.

Eventually, we reached a winding, potholed driveway swimming in puddles that led us to the gated entrance of the property.

'Have you been experiencing a great deal of rain here? We had shocking downpours late last week.' My comment as we both stepped from our cars.

'No, not at all.'

'It's waterlogged, the drive.'

'Oh, that's something else. Let's go.'

Cream and serene, three storeys high, partially hidden behind century-old plane trees, the house welcomed us with its principal door and many windows open. Outside, set within early leaf shade were Lloyd Loom garden chairs and a small, round table with a glass top. The guardian or housekeeper had set a perfect scene.

'It's a pity you are here in this season. Another two months and the gardens will be a blaze of colours.'

To the right was a large rectangular lavender bed quadrangled by a low hedgerow of rosemary bushes. There was something rather peculiar about its layout but I could not figure out what. We had parked beneath one of two towering planes. One stood sentry each side of the wooden gates. To the left was a monumental pond where carp almost the size of baby sharks were ploughing through still green water.

'Unusual, to find a pond of that size in the front courtyard,' I remarked, but the agent was not listening. He seemed to be in rather a hurry. I trundled off after him.

I have always loved rambling expanses of untidy land with paths that disappear and reappear without logic and this Napoleon-style early nineteenth-century property seemed to have been designed to tease with its secrets. All that it offered was, at first, immediately enchanting. It was a beautifully proportioned house, far lovelier and more graceful than our farmhouse. It carried more history and some exceptional period features about it and, importantly, it contained an ample supply of bedrooms for the two girls, their partners, their five children and any other offspring that might still arrive. Before visiting the interior, the Spaniard suggested we tour the grounds. He led the way. Beyond an oasis of greenery, paddocks and meadows, we came upon a sizeable lake fed by a distant stream and he confessed that as a boy he had spent happy hours boating on this expanse of water. Although he was Spanish, he had grown up in one of the neighbouring villages. His First Holy Communion and later Confirmation had both been celebrated in the chapel that stood to the right of the main house. He walked me to it, opening its heavy wooden door with a massive key. A cool, limed white interior with a dozen or so chairs, askew, furnishing the space.

'Masses are still held here.'

'When the family is in residence?'

'Erm, well, once a year, at least. The celebration of one annual mass is obligatory if it is to keep its consecrated status. One of the local priests, a fine father from Beaumes-de-Venise, performs the service. Of course, he has his own parish. His time is spoken for but he will occasionally conduct a wedding mass here.'

I pictured the young of the champagne dynasty tying the knot on perfect summer afternoons, blessed in vintage bubbles and advantage. 'How gloriously privileged.'

He harrumphed. 'Erm, usually, it's village folk, but the services for the local children, such as I knew them, well, officially, they have all come to an end.'

He did not explain for what reason.

I caught sight of a golden Labrador sleeping, spreading itself in the spring sunlight. He lifted his head at our approach but bothered us no further. I could hear a machine running. It was a gardener working somewhere close by.

'You should meet him. He's served this property all his life.'

'Ah, I wondered about the dog.'

'No, she's not his. The dog belongs with the house. She's very old. Too old to travel with the family, so the mistress of the place just leaves her here.'

How sad I found it just to abandon the poor beast.

'Yes, she's a hateful woman,' snarled the Spaniard. I was a little taken aback. Such information struck me as rather inappropriate. 'She fights with everyone. The entire town hates her.'

We had moved on; a pebbled yard offered a fountain decorated with a lion's head, snarling open jaws, through which the water flowed. Now we were examining another fountain, a hidden one, buried within the lower limbs of trees and bolstered by a makeshift rockery that struck me as quite recent.

'Ingenious, isn't it?'

There was a story that accompanied this dark corner – was it from the past or recent? I could not get to the bottom of it – but the agent seemed uncertain of its details, or unwilling to disgorge them.

'Where is this water destined, replenishing the carp pond?' It had struck me as rather stagnant.

'No, not the pond but its flow has created a rather successful reroute, I believe.' I could have sworn he was rather smug about it. His body language suggested a bird fluffing up its feathers.

I rose from my haunches, lifting myself up out of the roots. There was not an olive tree to be seen anywhere though we were only two hours from the coast. 'Is nothing grown, produced on the land?'

He shook his head. 'No, it's a holiday home, occupied for two weeks only a year. Such a waste. They descend from Champagne like an army, a massive entourage. Sixteen or eighteen of them, plus children and servants who sleep on the top floor. Otherwise, throughout the rest of the year it stands empty.'

'Any reason why they are selling?' I doubted that even in these difficult times one of the largest champagne dynasties in France was short of cash.

'I think she's fed up fighting, too many court cases. Oh, but that is all over now.'

'What is?'

My companion quickly realised that he had disclosed too much and would not be drawn, would not elucidate, fobbing me off with inconsequentials.

We passed on to the swimming pool. It had been sunk at the far corner of a field, out of the way. It was an odd, soulless addition, lacking changing rooms or a barbecue terrace or bowery where cool drinks might be served. Hospitality, *joie de vivre*, were absent. So, too, the kitchen when we eventually stepped inside the house. This was certainly not a room that represented the activity required to feed a household of twelve bedrooms.

'It needs a little freshening up, refurbishment.' Standard agent-speak. 'They never eat in,' he continued.

I also found it odd that the kitchen was at the front of the house, while the central staircase to the floors above was tucked away by the back door. Its layout was unusual in a property of this period and stature. I was beginning to realise that there was something curious about the building with its library in oak and its beautiful chestnut wood floors. I loved secrets, hidden gardens, but the house itself struck me as though woven in some web of deceit. I could not quite put my finger on what was odd. I had been told that the estate was 'competitively priced for the present market'. In fact, for the acreage and number of bedrooms, it was exceedingly reasonable, but for what reason?

After three hours, the truth had still not come out.

My escort led me to the neighbouring vineyard ringed by the mountains of lace, *les Dentelles*.

'The Dentelles range extends for fifteen kilometres and contains some of the finest walking in Provence, good paths, very accessible. Wild almonds pink the hillsides in early spring, followed by broom blossoms closer to summer. Tranquil countryside. You see how craggy and ragged those peaks are?'

I nodded, gazing towards sharp, teeth-like summits clawing towards a blue sky.

'Those are the result of a Jurassic trauma here. The limestone was forced upwards out of the earth and, over millennia, wind and weather erosion have needled away at them. And the views from up there, the nature, well, it is a spectacular spot.'

He spoke with an enthusiasm that suggested he loved this region, or was he also employed by the tourist board? He puzzled me, this man.

The vineyard itself was a vast, flat field of fifty thousand square metres nurturing twenty-five thousand vines.

'This plain – magnificent, eh? – is not a part of the *domaine*, but the vineyard owners would be very happy to cede it to any potential purchaser if it assisted with the sale.'

'Cede it?'

'Cede, sell,' he shrugged, Provençal fashion. 'An amicable previously established settlement. The woman who owns it paid over the odds for it twenty-five years ago, but I know she would agree to a cash sum just so long as the contract stated the legal minimum. She doesn't want to get saddled with state taxes. And, consider, with this vineyard in the estate papers, the newcomers could register themselves as agriculturalists and thus negate all annual land tax.'

I was bemused. We continued to pay land taxes, but this was another *département* and it was possible that the statutes differed here.

'You could dig up the vines and plant olive trees. And the water issues would mean nothing with hardy, drought-resistant olive trees in place.'

'Which water issues are you speaking of?'

He eyed me beadily. 'Olive trees need precious little water,' he replied testily. 'And the region boasts an AOC for its olives and oil.'

'Yes, I know. Are farmers troubled by invasions of the *oleae* fly?'

'Unfortunately, yes, but there are one or two excellent pesticides that deal with the problem.'

'Well, then, no olive trees for me,' I muttered.

The wine was an average table offering and, no matter who farmed it, it could never be much improved upon, but the district produced many fine wines.

'Gigondas, for example: what exceptional reds! But this particular vineyard renders wine that is drinkable, nothing more. It sells at three euros eighty-five cents a bottle. Currently, seventy thousand litres a year is bottled. So, there's an interesting business concern here. I don't know the sum Thierry suggested for the vineyard, but I know for certain the vendors are open to offers, or if you don't want wines, go for olives. As I said, precious little water necessary.'

For an estate agent, he was beginning to sound a little desperate, and the water issues were baffling given that he had assured me the *domaine* and surrounding lands were very well serviced by subterranean springs.

At the edge of the vineyard was an eighteenth-century *lavoir*.

'This could be pulled down,' suggested my guide.

'Oh, but of course not!' It was an attractive, authentic feature. There were several natural springs that fed into this parcel and their routes down from the mountains were being pointed out to me when, suddenly, the Spaniard noticed that the path alongside the vines was waterlogged, as the entrance-way had been. Water was seeping from somewhere in the centre of the vineyard. He was on the phone instantly, hot and bothered, shouting to someone about never having seen such a sight here and what was going on.

'It's bled off into the vineyard,' he was hollering. His Spanish accent with its thick, lispy sounds became evident when he grew angry. I was curious as to why he was personally so worked up about the situation. He surely did not think that such an incident would prejudice this potential sale. Or was there a personal issue at stake here that I was not party to? Fishy was the adjective that sprang to mind.

Returning to our cars, I noticed just beyond a bank of exceedingly tall trees to the left of the property, beyond a stone wall that must have measured ten metres in height, along the only side to the house we had not visited, another house, older, possibly medieval. It must overshadow the windows on the west of the property. Why had I not noticed it? Because the shutters had been closed on that side of the house.

'What is that building?'

'The original house.'

'Is it being sold with the estate?' As we drew closer I saw that in its stylistically different and original way it was substantial.

'No, there's someone living there.'

Oh. I pictured all privacy disappearing. 'Ah, that's the troublesome neighbour?'

'No, no, he's extremely kind, but is terminally ill, I understand. Not improved by the years of acrimony and fighting that has gone on with *Madame des Bulles* here.'

The continued allusions were too frequent now to ignore

and I was keen to hear the story. So I stood my ground and insisted and, reluctantly, he began to spill the beans.

The tale proved to be a real-life Pagnol affair. Water, then, was at the root of it.

Originally, the estate had comprised both houses and all the lands. The main entrance was elsewhere and belonged now exclusively to the expiring neighbour. However, all water sources had by an oversight or trickery at the time of division of the properties been left with the Napoleonic chateau. This forced the original house and the vine groves to be dependent upon the chateau for water.

'Here, in this part of Provence, you understand, water must be shared, particularly from springs.'

However, soon after her purchase of the chateau more than two decades back, Madame from Champagne had closed all access to the wells and sources. Wherever a stream or underground rivulet ran beyond her land she had blocked it off, thus sequestering all water for the chateau's exclusive use. Incensed, outraged, the local population as well as many of the outlying farms became involved. The neighbour, younger and healthier at that stage, had, as a retributive act, closed off the gated entrance that had served both properties by permanently fencing off from the chateau the lane that ran the length of the main park. It was his legal right to do with this approach as he pleased, but the sharing had been a gesture of goodwill expected of '*voisins*'. Now it had been withdrawn. War had been declared. His act had forced the Champagne family, les Champagnoises, as these men had delightfully nicknamed them, to enter by the 'back passage', the servants' and merchants' access. This, of course, had infuriated *Madame des Bulles*, the Lady of the Bubbles. She had taken the man to court and he, in turn, had brought a suit against her for water rights, as had the owners of the vineyard which, it turned out, belonged to the in-laws of the Spaniard.

How could I have been so shortsighted? The lavender beds had once been a tennis court – 'Well, you can hardly have a tennis court in the forecourt,' grinned the man at my side – the carp pond, once in an area of tranquillity at the back of the house by the kitchens where coffee might have been taken in the mornings or where the staff would have collected herbs and fed fowl.

The house had been reversed. What had originally been conceived as the back door was now the front entrance and vice versa. The splendid sweeping staircase now at the rear of the ground floor had been a greeting point when the property had been entered by its original front door.

'Her pigheaded foreignness' had cost the woman from the north dearly. The community had closed ranks against this house.

We were now in the company of the gardener, whose name was Baptiste. Tall, rather handsome with healthy, rugged skin and white hair, he was leaning, propped against a straw broom. He had long since given in his notice and only dropped by during the proprietor's absence because he did not want to see a property so fine go to rack and ruin and because he grew all his vegetables here.

'But who feeds the dog?' I begged to know.

An insignificant clerk from the local town, without wife or family, had been given sleeping quarters in a lean-to hidden behind stands of bamboo near the chapel in return for that service.

Baptiste, along with dozens of others from the neighbouring villages, came to help themselves to the fish in the carp pond and to the seasonal fruits and flowers. Occasionally they used the chapel, too, but 'she' knew nothing of any of this. As the agent had said, from time to time a marriage was celebrated and the fine glassware and crockery, as well as some of the house's exquisite furnishings, were brought out into the garden to service the wedding parties.

'If we considered purchasing the *domaine* might you agree to stay?' I asked Baptiste.

He shook his head. The place carried bad memories for everyone. 'You will find no one who will work here. In one sense, we judge these grounds cursed. On the other hand, our sentiments are intermingled with our past histories here, our childhoods and that of our parents and grandparents. Today, we simply take advantage of these grounds and most particularly the water sources.'

Everybody came here, walking with buckets or driving up the back yard, to avail themselves to the spring waters that 'she, the foreigner' had closed off from them and their neighbours. From every direction, paths and tracks led the Montmirail residents here. The hunters knew how to penetrate her forests, to hunt the boar and the rabbits and hares, the young girls came to filch the flowers for their garlands. Everybody helped themselves to water and, when she was not in residence, many of the taps were opened and left to run freely, to irrigate every outlying farm, olive grove and vineyard at her expense.

'Is she aware of any of this?'

The two men exchanged glances and then the gardener, with his iris-blue eyes in a face that watched out at the world, quietly calibrating Nature's shifts of mood, replied that he doubted whether she would have noticed. 'Her thoughts and concerns lie elsewhere. She sees very little, though she deems herself smart. But we are smarter.'

And the court cases, the litigations, had they been resolved?

'More or less.' Hardly sufficient reassurance for a potential purchaser.

Well, then, might there be any chance of a brief interview with the neighbour, to understand his point of view regarding these perturbing and not entirely resolved issues? Both men solemnly shook their heads. 'Oh, no, he lives alone at death's door, waiting for the final call. And if this becomes an inheritance issue ...' the men grinned. 'She'll never unravel it. If you want it, buy now before he dies, before it is too late.'

Otherwise, the chateau would continue along this rudderless present, a diminished asset, repossessed in spirit by the progeny, the descendants of the local peasantry whose ancestors had toiled its expansive lands.

I liked the house immensely, was drawn to its elegant space and generosity of bedrooms, but would we not be purchasing as many problems as I was trying to solve? The fly back home was the devil I knew. As I turned the car towards Avignon, I tapped out the number of Vincent's mobile.

I rarely found the opportunity to visit the papal city of Avignon but this seemed to be the perfect opportunity. Disappointingly, Vincent, from the Agrivert team, was not working. He was, however, in town and agreed after a little persuasion to meet up with me briefly. His location of choice was not his office, but a café outside the enclosed walls of the historic city centre. I spotted him as I parked – I'd got snarled up on the circular – looking tired, pale. Instead of coffee, he ordered a Calvados. I took Badoit. I had a long drive ahead of me in early evening traffic. I was hoping Vincent might clarify the situation on products awarded organic certification.

'The farm we visited where you had planted up the wild flowers, the owners, manager, are using a spray created by a top-tier American crop-management company?'

'An *attirant*,' he said. 'It attracts the fly to it and then kills it off.'

'But it attracts and kills other insects, too?'

'Yes.'

'We have kept hives on our land and I want to do so again. Would this be a risk for them?'

'Certainly.'

'Then how has this product received an organic ticket?'

'Its basic composition is from a fermenting plant so it is classed as a natural substance, not a chemical one.'

'But it is toxic for the honeybee and other insects?'

'Yes.'

'So, why had it been given an organic ticket?'

'In my opinion, it is hazardous, though, to be fair, it is not on the Dangerous Pesticides list, so compared to dimethoate, for example, it is the softer option. *Bio* is a growing movement. Awareness is on the increase, more and more is being written about the hormonal imbalances, cancers, side effects, you name it, caused by agrochemical products and the government has been picking up on this. Many people are worried by talk of poisons being fed into their systems. And these are urgent concerns. This product is technically speaking organic because it is plant-based. It will help kick-start the organic olive industry, which, as you know, until we find out the results of *Psyttalia* is an almost insoluble problem. It gives olive farming the organic start it so badly needs.' He shook his head and downed in one the Calvados set before him.

I reminded him of the pyrethrum chrysanthemum. Why would this American product, which was also non-selective in what it killed, be given a ticket of approval while the Dalmatian plant was not available?

'It's a compromise. There are many of us working in the *bio* area who are not happy about that.'

What other directions did he feel the *bio* path should be taking? What were his thoughts on *Psyttalia lounsburyi*?

'It might prove to be a way forward in the future, but for the present our greatest hope lies in healing the earth. Little by little.'

And in order to achieve that, in Vincent's opinion, we needed to change our behaviour patterns immediately. To turn our backs on philosophies that supported *taking* from the land and giving nothing back, that applauded bigger and better crops.

'We are told that mass crop production is essential to serve the phenomenal population explosions, but have such methods curtailed the rising levels of starvation on the planet? No. The truth is, man is confused and what we are fighting against is immense: international financial interests that are greedy, corrupt. In my opinion, we must be very wary of Monsanto, Dow, Bayer, individual governments and the self-interest displayed from within them. Do you think any one of them gives a hoot about the future health of the planet? It's down to us. Every one of us. My job, as much as anything else, is to expand people's awareness of Nature, of landscape. Urban wasteland, for example, that is our landscape, our future heritage, just as much as the mountain ranges, the olive groves, the vineyards. Let's encourage city kids to farm those plots. Even if all one can contribute is one window box, one barrel in a dingy yard planted with flowers or vegetables, it's a beginning. It's the first step.'

7

Our Herculean pruning programme was almost at an end. Those trees that had not yet been touched must now be left until the following year because, on every branch of our silver soldiers, buds were beginning to peep through. That brief season of silky grey promise was upon us. The white blossoms were soon to come, shrouded for now within nodes barely larger than pinheads. Beyond buds were flowers and then came fruits and, as they plumped out, decisions . . .

Weeds, land growth, plants of the *maquis*, were shooting up fast. The frogs were mating. Soft rain was falling. Michel was soundly sleeping, long hours, bushed, worn out from labouring on the farm. I was suffering from insomnia again. Silently torn by the prospects that lay ahead during the summer months, I had been encouraging all this heavy pruning, working at it myself, arm-aching physical exertion as much as I was able, aware that the more rigorously we pruned, the less substantial would be our fruit load and the less urgent would be the knotty question of whether we sprayed or not. The alternatives that I had come up with so far were all unsuitable as I continued to dream unrealistically of a world without Dacus.

Michel found me seated in one of the armchairs, listening to the birds outside, the pre-dawn chorus, a sure sign that the days were growing warmer and longer.

'How would you like to spend your birthday?' I asked him.

He burst out laughing. 'Is it giving you sleepless nights? Something intimate, just the two of us.' He rested his hand on my head and urged me back to bed. 'Staying awake won't resolve anything,' he mumbled, padding back to the reassurance of pillows, bedlinen, oblivious to the fact that I was not accompanying him.

Morning broke and I was still there. At my side was a collection of dictionaries and botanical encyclopaedias. I had frequently asked myself why certain flowers, plants, were acceptable and others not. Where did the word 'weed' originate?

Here, offered by *The Oxford Dictionary* was what I found for WEED:

1. A herbaceous plant not valued for use or beauty, growing wild and rank, and regarded as cumbering the ground or hindering the growth of superior vegetation
   b) a plant that grows wild in fresh or salt water
   c) used with defining word to form the names of wild plants such as BindWEED, DuckWEED etc.
2. Any herb or small plant
3. Tobacco
   b) a cigar or cheroot
4. An unprofitable, troublesome noxious growth.

At what stage had some of the glorious plants, wild flowers some, others such as burdock, been cast to one side and labelled 'weed'? Botanical racism!

In certain cultures and kitchens, burdock roots were highly prized. The stinging nettle was another. Nettle soup is a delicacy and a great bonus for good health. High in iron, it counters anaemia and hair loss and is the exclusive larval food plant for several butterfly species. I was interested to learn that *The Oxford Dictionary* suggested that a weed is an 'unprofitable, troublesome noxious growth'. This rather damning and dramatic interpretation was the sense in which it was commonly used in modern parlance rather than as 'any herb or small plant'. When did that shift in perspective take place, and in what ways had it biased our relationship to gardens, to the plants growing around us? Weeds, creepie-crawlies, bugs. They had become our enemies. They turned us against Nature, encouraged the perception that we had to keep its growth, its profusion, at bay, before it took over, got in the way. Was not such a philosophy the basis upon which the giant chemical companies were proselytising? 'Kill off all weeds, ground cover, bugs, and the irrigation will be fed directly into the crop.' A philosophy put forward by industrialised and monoculture farming.

The sun was shining. Higher up the hillside, as I returned from the gate with the post the Portuguese were rowdy and cheerful. Earlier in the morning their smoke-ridden lorry had taxied up, delivering a second cement mixer. I noticed Quashia, who was working with sand, cement bags and spade, coveting it. It seemed that wherever I walked on the land cement was being mixed.

My email to the States had met with silence, and I was beginning to believe that it had been a prank, Spam. Instead, yet another envelope from the olive bodies had plopped into our letterbox: our grant, our stipend from Brussels, miserly though it was, a mere four hundred plus euros per annum, had been withdrawn. The notification informed us that no further transfers would be made into my bank account. The reasons given were that we had delivered no *récolte*, no harvest, the year before.

'But the latest European legislation states we are entitled to that aid whether we harvested fruits or not. Remember the tobacco farmer I told you about in Extramadura?' I argued.

M raised his hand to silence me. There was more. Last year, according to the letter, our forms had been inaccurately completed. Neither of us understood what inaccuracies they had found or whether these were the real reasons for penalising us, but for such an insignificant sum it did not seem worth the hassle of finding out.

'I'll see if I can get to the bottom of it. Otherwise, if it proves to be too complicated, we'll just let it go,' sighed Michel wearily.

He filed this letter along with the others from the olive farmers' mutuality, from whose membership we had also been expunged in spite of numerous attempts on his part to persuade them to the contrary. Slowly, we were slipping off all lists, all those societies and bodies Michel had gone to such lengths to have the farm registered with.

'Does it really matter?' I asked him over dinner that evening. Bureaucracy leaves me exhausted.

'If we continue to produce no fruits, press no olives, then we will eventually lose our right to an AOC rating. It has taken us a very long time to achieve all this, Carol. Consider our options. Luke delivered no predator, which in any case has been judged ineffectual, and the one "organic" alternative seems to be fatal for bees. *Chérie*, I was behind this transition but if there is no possibility of making it work, what do you want us to do? Give up on olive farming altogether? And then what? Let the fruits rot and live here like tourists?'

Here lay the problem.

'We need larger premises if we want your daughters—'

Michel shook his head. 'We are not moving. We are laying the groundwork now for an extension.'

'Then, I don't know.'

'What would you say if we chose to continue with one of the chemicals for the time being – wait, before you argue! – choosing the most benign, the least offensive, and we work with the agricultural organisations, taking a responsible and active role in pushing this issue forward? The trees are budding. We have to make a decision.'

His words were fact. A decision had to be settled on within the coming weeks. I simply did not know. My instinct was to let the fruits drop, but on the other hand it struck me as a shocking waste of an excellent comestible product. I was caught between a rock and a hard place and I did not know where to turn to for advice.

'Perhaps you have a point,' I conceded though I did not believe my own words. 'Maybe we will have to spray just one more time.'

'We can try the American product,' smiled Michel.

I nodded despondently.

Serendipity had been a major player in my life at Appassionata. Once the estate had been purchased and the olive trees had revealed themselves, René appeared, a chance encounter, and offered to teach us how to husband them; Quashia walked into our lives when I most needed him. On my travels circumnavigating the Mediterranean, the discovery of the six thousand-year-old Lebanese trees had been a masterful stroke of good fortune, setting me off on journeys that had changed me, my perception of the world . . .

Yet, now, I was ready to stand defeated while the land was treated with products I strongly opposed. We were set for a good harvest, judging by the quantity of buds on the trees, but I had failed to find an alternative method of protecting them. Michel informed Quashia the following morning that we had reconsidered our position; we would be cleaning up the spraying machine ready for its summer work. From the balcony above, I saw the smile burst across our old Arab's face.

I returned to my desk, unable to juggle the choices any longer. The chateau near Avignon with its convoluted water issues might not be our answer, but that was not the only property available. I booted up my computer ready to go to work, to re-immerse myself in my travels, in the sandy deserts of elsewhere. And there it was: a response from the States.

The email read: *Our technical people will be in Sophia Antipolis at our offices this coming Wednesday. They would be happy to meet with you and to answer any questions you might have about the mouche d'olive in general as well as specifically about our product.*

Yours.

This was the firm that had patented the organic product I, along with others, deemed dangerous to bees.

I tapped an answer immediately. *Please, send me the address of your offices and a time of appointment. I will be there.*

Gentle spring rain fell overnight. Still, the Portuguese were cracking on in spite of the long lunches and the crates of emptied wine bottles. A sight to behold. A trio of them, all the Josés seated on our upturned dustbins, secured with a plank, at a small garden table covered over with a discarded wooden door to create extra length. Bottles and food aplenty, the dogs forever at their heels.

'Please don't feed the dogs.' My daily mantra repeated later from the balcony.

'*Si, si, Madame*', pouring yet another litre of red wine into three plastic goblets.

From the other side of the cypresses, up a ladder, climbing about within the canopy of one of the elderly olive trees, was Quashia, completing the last of the giants for this year, throwing disapproving glances towards the masons.

On my way to the washing line, I spied our two Alsatians

chewing contentedly. 'What are they eating?' I called to anyone who would listen.

'Just a small portion of *fromage*, Madame.'

'But I asked you, please, not to feed them.'

'It was nothing, just the remainder of the cheese.'

I had failed to make these men understand that feeding the dogs encouraged the beasts in their belief that every comestible mouthful on the estate was theirs by divine right.

My meeting with the two specialists from the American agrochemicals company was to be held at their regional offices, spitting distance inland from our property. They were situated on the outskirts of Sophia Antipolis, a city constructed *ex nihilo* in the heart of what had been sprawling forestlands of cork, evergreen oaks and Mediterranean pines. It had been conceived and hailed as France's answer to Silicon Valley. Both were situated in similar close-to-coast locations, but the French version was on a more modest scale.

Its nomenclature was derived from Sophia, the Greek word for wisdom, and Antipolis, the original name for the trading harbour founded by the Greeks somewhere around 500 BC and annexed by the Romans in 43 BC, what today is Antibes. One of the websites for Sophia, as it is known down here, claims it to be probably the 'most cosmopolitan spot in the whole of France' with thirty thousand inhabitants gathered from sixty countries on twenty square kilometres. The companies based here, predominantly American (English is the lingua franca in Sophia not French), tend to be working in research and development, specialising in technology: computing, electronics, and pharmacology and biotechnology. The place is also a telecoms hub. This was the high-tech Côte d'Azur, corporate Europe *par excellence*, stowed within a maze of science parks and endlessly circling, figure-of-eight roads that seemed to lead back to the same manicured roundabout planted with olive trees and colourful flowers. In fact, there are a number of these roundabouts within its radius and they all looked identical.

I set off early, ridiculously so, because I always found myself lost whenever I ventured into this curious, rather troubling zone of our coastland nestling within the silent but not necessarily acquiescing pine forests. My meeting had been fixed for 6 p.m., after the two men had completed their day's work. They were 'fitting me in', staying late to meet me, I had been informed, and I did not wish to keep them waiting or to lose out on this unlikely and rarely granted audience. I crawled the roads slowly, impatient cars edging up my backside, true road manners of the French, obliging me to pull over to the gutter every few minutes to let them pass. The journey took me fifteen minutes because their beige, smooth-stoned complex of buildings turned out to be a short distance outside the techno city's central hub. Surprisingly, there was no gated entrance, no guard control to quiz me before I descended a steep asphalt lane and swung left into a car park where each space was numbered, reserved.

One did not just happen here, did not just drop by. You drew up here because you had business with the international firm whose logo, as tall as any man, stood resolutely in a grass patch outside the doors of one of two separate buildings.

It was the end of the working day. I slid into bay fifty-two, the first free slot, switched off the engine and drew out a book, but I was idling, not really reading, feeling unnecessarily anxious.

The park's vibe rattled me. It was still, too still, until a man with a red paisley bandanna tied round his forehead, Corsican pirate-style, materialised from the surrounding pine woods, puffing, exercising himself and a big black-and-tan dog. He was followed shortly after from the opposite direction by another fellow, grey-skinned in floppy grey shorts. This one jogged the pavement past the bonnet of my car, but did not register my presence. These corporate employees, their exercising, struck me as incongruously ordinary for this patch of techno parkland. Yet the denizens of Sophia lived hereabouts. Not in these neutral blocks outside of which I was marking time, but elsewhere, secluded within the forests, in gated, guarded and alarmed villas and apartment house complexes. In fact, no other single influx of foreigners aside from the new Russians, with their bootloads of illegally earned dollar bills, had contributed to the healthy upsurge in the property market down here as effectively as the multi-national dollar that financed the white-collar expats who had swarmed into Sophia and populated this backwater, occasionally descending to the coast for leisure activities such as sailing, golf and expensive meals.

At a quarter to six, unable to sit composedly any longer, I stepped out of the car, flicked the lock and strode in through the sliding doors of *Bâtiment 1* to reception, on the ground floor at the rear of the building. A young, dark-haired woman behind a counter greeted me, requesting my name and business. She looked semi-deranged with a slick of orange lipstick that had escaped her lower lip and smeared her chin, with hair drawn into a chignon falling untidily about her face.

'Business card, please?'

I shook my head. 'Sorry, I don't have one.'

This puzzled her. 'Appointment?'

I nodded. Behind her, a pair of glass doors were open with a muslin curtain floating lightly in the wind. There was something ephemeral about it, ghostly, quite out of keeping with the hard-nose business that the occupants of this place were involved in.

One of the company's promises: Helping Growers Sustain the Soil.

'Well, who with?'

I reeled off the names of the two directors. She seemed perplexed, had never heard of them, picked up the phone and dialled an extension, asking what she should do. Had I made a mistake? At that moment, a middle-aged man was at my side. From where he had come, I had no idea. His arrival had been as silent, as stealthy as that of a cat burglar.

'You must be Madame Drinkwater?'

I nodded.

'You are early.'

I found myself apologising, offering to wait on the seat close by the open door that I could now see led to nowhere but a high stone wall. Dead end, I thought.

He smiled a smile that I could only describe as American although he was French. It was full of even white teeth and signalled assurance, ease, success. It was a well-practised smile, lighting up steel-blue eyes and metal-grey hair, neatly barbered. 'I'll just collect my briefcase and then we can make our way to the *salle de réunion*.' He was gone and back in an instant, carrying a black leather bag the size of hand luggage. He escorted me to the end of a corridor and we turned left into a conference room furnished with an oval wooden table about six metres in length with fourteen chairs spaced around it. Spewing out of its centre was a spaghetti of wires, a pair of microphones, leads for broadband connections to laptops and two glass jugs of cloudy water. At the end of the room, dominating like a schoolteacher, was a blank rolled-down screen.

He pulled out a chair for me, but without thinking I chose another and he settled on the one he had already drawn out. His pose was relaxed. He leaned backwards into the seat, one leg crossed over the other, in a wide-open manner. He wore a light blue cashmere sweater slung over his shoulders and loosely knotted over his darker blue linen jacket. Immaculately pressed slacks led the eye to expensive loafers and diamond-patterned Burlington socks (of the sort worn in certain Parisian circles as a trademark of class, of aristocracy, of leadership).

'Let us take the time to introduce ourselves while we wait

for my colleague, who, like me, is also christened Jean-Christophe. He will be with us shortly.' He gave me a very brief introduction to himself, saying that he was the boss in France of the technical division of the company specialising in agricultural sciences and that he was not only responsible for the marketed products but in certain instances had created them. 'Now, tell me who you are.'

I was surprised that he did not know, given that his firm had contacted me in the first place. He shrugged dismissively, saying that he and his associate, the second Jean-Christophe, who had just telephoned to say that he was on his way down, had received a message from head office in the States via a technical director in Spain, requesting that they meet me and assist me in every way possible with products for my olive farm.

'I have no idea who you are,' he smiled.

I gave what I judged to be a truthful assessment of myself and finished with the fact that I wanted to be frank, that I was looking for an organic way forward, not a path that was pesticide-driven.

There was just a fleeting hint of mocking amusement in his expression.

At that instant, Jean-Christophe Two appeared. (I smiled silently to myself, recalling the three Josés.) Marginally younger, open-neck white shirt, no jacket, black Armani jeans, expensive loafers. He, too, possessed one of those wide white smiles with a full deck of teeth beneath a healthy shock of dark hair. He was a little podgy. We shook hands. He sat alongside his colleague and the two men grinned at one another and then at me.

'You haven't missed anything. Madame Drinkwater and I were just doing the introductions. She's looking for organic products.' The way he spoke this last sentence, I felt it might have been a code.

The technical director, Jean-Christophe One, laid on the

table a small brochure of the product that had brought them organic recognition here in France. 'If you are looking for organic, here's your answer,' he said, pushing the leaflet in my direction across the polished surface. He was trumping, playing his winning card.

This was the product that had been talked about by the Agrivert team I had spent the afternoon with a few weeks back, the very same spray that Vincent had admitted to me, when we were alone in the café in Avignon, had been given its ticket as a trade-off, a compromise.

I picked up the leaflet and studied it. It showed a rather gruesome series of photographs detailing the detrimental effects on olives caused by Dacus, our destructive fly, in essence a fruit fly like any other, which lays its eggs in the pulp of the growing fruit. It displayed in graphic detail the cycle of the fly's life. The female rises up out of the earth and lays her eggs, anything up to four or five hundred in as many olives. After an incubation period of two to four days, those eggs hatch into larvae. Thirty to ninety days later, the next generation of fly emerges. This process of egg-laying and birth takes place three to four times a year. Four generations of flies emerge to propagate once more.

The men were watching me while throwing occasional glances, triumphant smiles, at one another.

I was studying the leaflet silently, staring at photographs that told a horror story. These images had been selected purposefully, chosen to represent the gruesome enemy of man at its most destructive. I am the least tactically driven person I know. I live by my emotions, my passions and all my decision-making comes from my heart, but I was fully aware that I had to tread carefully. I was in the company of two highly skilled individuals, trained to sell, to market and to ward off, iron out all doubts and concerns. Their company policy was about peddling solutions. They were committed to 'shaping the future of agriculture and pest management'.

Their products were deployed for use as 'weed, insect and pest management'.

It was not my place to humbly suggest that perhaps the world, the growing, producing earth, did not need to be shaped and managed.

What was to be gained by perceiving the natural world as an enemy?

The answer was, plenty.

I reminded myself silently, still perusing the leaflet, that, after Monsanto, theirs was one of the most influential agro-chemical firms on the planet and these two were power players within it. I wondered again why they or colleagues of theirs had offered to donate this time to me.

I began slowly, warily. I was nervous and quite unskilled in this game. I told them that another of their products, not the organic one the leaflet was describing, but also containing a base of Spinosad, had been proposed to me by their head-quarters, but when I had studied its properties carefully, I had found that it was highly toxic to bees. And I believed that this product, the one detailed on the leaflet, based on the same composition, was also dangerous for honeybees and without honeybees, I said, the world would be facing an economic as well as ecological disaster. The value of honeybee pollination to US agriculture alone was more than fourteen billion dollars. Orchards in which honeybee hives are kept *always* produce more fruit than those in which there are only visiting pollinators.

The older of the two, the technical expert, waved his arm, assuring me that the possible risk to bees was negligible.

'How do you arrive at that conclusion?' I asked.

'Honeybees don't frequent olive groves.'

'It is a fact that the olive flower is not pollinated by any bee, honey, bumble, feral, or a wasp. The tree is auto- or self-pollinating or pollinated by the wind. So, you are correct in saying that bees do not search out the olive tree's tiny white lace blossoms. However, in a healthy grove where wild flowers have lodged themselves and have not been wiped out by weed management products, where the grasses around the feet of the trees have not perished due to herbicides and where the soil is fertile, there are wild flowers in abundance, ergo there is pollen and nectar, sustenance for these endangered arthropods.'

The gentlemen before me dismissed this argument.

'Oh, so few as to make no difference' was their response. 'Certainly, if a bee happened along and was attracted by the globules of sprayed product – because this product is an attractant so that when the insect seeks it out and makes contact with it, it is poisoned – the bee would be killed almost instantly.'

'So, it does act as an attractant to honeybees?'

'Listen, our company,' chipped in the other firmly, 'is dedicated to the environment. The substance that this product is based upon comes from a plant, it has been created by a natural process of fermentation. This stuff is good for the farmers. We have spent a great deal of money getting this file through here in France. To register a product is exceedingly expensive and every country demands that it is registered with their own governing bodies before it can be sold in their territories. In terms of the olive market, France is peanuts for us. It has cost us more than we will recoup.' He flicked open his computer to confirm the hectares in France growing olive trees: 'less than two hundred thousand'. Compared to Italy or Spain, or indeed almost any other oil-producing Mediterranean country, the figure was very modest.

'France doesn't even make the top ten,' he grinned.

It is a fact that compared to Spain's 970,000 tons of olive oil produced every year, and climbing, France's output of approximately 2500 tons is insufficient even for its home market. It is one of the reasons why French olive oil is so hard to obtain outside France.

'Peanuts,' confirmed the technician. 'But,' he continued, 'we know that we have got the only real solution out there if you want to get away from the old-fashioned pesticides, those that are slowly being withdrawn from the marketplace—'

'—because it is now acknowledged that they are dangerous for our health and the environment?'

He wagged his head back and forth. 'Perhaps,' he allowed eventually.

'So, may I ask why you have gone to this trouble, to the expense of registering this product in France, if it is not financially viable for you here?'

'We are committed to the environment' was the swift response. 'We are just like everyone else, you, your family, your olive farming colleagues; we want to see a green future.'

'Are all your products organic?'

No, this was the only one with an organic ticket. They were marketing several others in France that were not organic. The others were their money-spinners. They were working with fruit arborists and *viticulteurs*. The latter, the wine industry, was, of course, a hefty business in this country. They did not mention wheat or tobacco.

'Might I suggest that registering a product that carries an organic ticket earns you pittance revenue but is valuable to you because it gives your company a renewed image in the eyes of the world?'

Along with the giant pesticide producer Monsanto, this company had also been a producer of Agent Orange, the chemical defoliant, the herbicide created from a base of dioxin, used by the US military during the Vietnam War. Between 1962 and 1971, seventy-seven million tons of this herbicide was sprayed from low-flying planes and helicopters into the jungles of South Vietnam to expose the Viet Cong, the enemy hiding within these leafy parcels. It was the key player in America's Herbicidal Warfare Program.

The seventy-seven million litres of this plant defoliant

sprayed into Vietnam have left residues of destruction that are still being logged. Even today, there exist high incidences of Vietnamese children born without limbs, blind, or stricken with unimaginable cancers. Almost inconceivable are the numbers of stillbirths, miscarriages and other defected newborns. I have travelled in Vietnam and I have met with some of the post-war victims. These figures were not anti-American propaganda, but I refrained from touching on any of this because these men represented a new generation of technicians. Jean-Christophe One, the man who had greeted me at reception, had been with the company for thirty-seven years he mentioned a little later in the conversation. (He certainly did not look old enough.) A swift calculation told me that he must have begun his employment with them in 1972 when the Agent Orange programme in Vietnam had just been called to a halt.

And what of tobacco sprays? Tobacco is a major crop in Europe. It garners more European funding than any other. I had learned this during my travels in Spain. It is very big business even while, on the one hand, governments are banning smoking in public places. Cigarette packets in Europe and the United States carry the warning that death can result from smoking. Everybody knows it, whether they act upon it or not. Tobacco is bad for your health.

But I had discovered a fact that I found equally disturbing. Pesticides, insecticides, in the modern understanding of the words, were developed in Italy during the 1820s for use on fruit crops. Those first products were created from a base of ore of arsenic. Even as far back as the 1850s, these chemicals were discovered to be carcinogenic, the mineral ore of arsenic being the culprit, but these sprays were not withdrawn. On the contrary. They are still being developed and today one of the biggest markets for ore of arsenic-based pesticides is the tobacco industry. The crop leaves are sprayed relentlessly with these products known to trigger cancers. Yet still they have not been withdrawn from the market. I have often asked myself whether it is the act of smoking that causes the cancers or the poisons sprayed on to the growing plants to kill off the pests. Perhaps a combination? I put this to these two men and their response was: 'Our company doesn't produce a product for tobacco crops. Smoking is bad for you.'

Neither of them smoked.

Residue of arsenic has been detected in groundwater.

What could they tell me about their products and seepage into groundwater?

The leaflet on the table was pushed a little closer to me and then Jean-Christophe One picked it up and turned it over, pointing to a small box. 'Here,' he indicated, 'it tells clearly at what distance from water this product may be used. Five metres is all in the case of this organic pesticide, which proves how safe it is.'

And then the direction of the conversation changed, skilfully rerouted. Both men wanted to talk to me about what 'organic' really meant.

'You say, you want to run your "little farm organically". But do you understand what that actually means? You know, we're working in this business, we see what is going on out there.'

'I am keen to hear what is going on out there.'

An example, they said. They had eaten together in a very fine restaurant in Montpellier a few weeks earlier and they had ordered bottled water.

'It was organic,' the dark-haired Jean-Christophe informed me. 'And you know where it came from?'

I shook my head.

'Organic water in a glass bottle from your country, Ireland. Think of the *bilan carbone*,' he said. 'How ridiculous is that!'

And they both grinned their expensive smiles, enjoying the joke.

I had to think for a moment. *Bilan carbone*? Ah, yes, carbon footprint. Bringing in water from Ireland, the expense of it, the glass bottling . . . yes, I agreed, it seemed totally unnecessary when France produced very fine mineral water, but what had stopped me in my tracks was the fact that I had not told them I was Irish, and they had not known anything about me, they had both claimed.

Now I was being regaled with horror stories about the organic and non-organic food markets. Each worse than the last and each narrated to prove to me how squeaky clean their position, the integrity of their company, was, that no such dirty tricks were being played by them.

I knew that I was out of my depth. Whatever I said or asked, there would be well-constructed counter-arguments.

'What do you have to say about the world's rising cancer figures?' I had been reading that very morning an article published in the weekly agricultural journal I subscribed to that the use of chemicals on the land and plants, along with the chemicals used in our daily lives – washing-up liquids, detergents, plastics, paints, but, above all else, the products being ploughed into crops, and inevitably the soil and water sources in which the crops are existing – had caused cancer figures in France to rise over the last twenty years by shocking proportions.

'Leukaemia in children,' I cited, 'is on the increase every year. Breast cancer has doubled and prostate has tripled.'

'We are living longer, stuff happens' was the response. Feeble, by any standards.

How did they feel about the fact that in a world that was waking up to its environment, the pesticide giants were perceived as the enemy?

Number One replied that when he shaved in the morning he could look himself directly in the eye. He had nothing to feel responsible about. The second grinned and said, 'Look, I am not knowingly trying to kill anyone.'

Both agreed that because they were working and living

within the framework of a chemical giant, they knew the truth, they knew what was *really* going on out there. There were horror stories, sure, but not of their making. Did I know anything about the pig market?

Nothing at all.

A large proportion of the pork foodstuffs imported into France – where the growing of genetically modified crops had been halted while the subject was studied – had come from pigs that had been fed abroad with genetically modified soya-based feed.

Monsanto – they did not mention the company's name – owned and grew Roundup-resistant soya bean plants world-wide. Roundup is the world's biggest selling pesticide. Their soya plants have been genetically modified to resist Roundup so that their product can be sprayed in vast quantities in their soya fields, killing off every blade of grass, insect, microbe, fungus and 'weed' while the crop itself survives, unaffected.

Recently published data enumerating laboratory test results of genetically modified foodstuffs on human/animal health was deeply worrying to say the least and what is even more horrifying is that Monsanto plays dirty. It is publicly known that they have placed their own people in positions within the FDA (the American Food and Drug Administration), to ease the passage of their products into the marketplace. It has been suggested that they have forced the voting and vetoing powers within other governments, restraining scientifically studied information, to achieve the same results abroad as at home in the States.

I had little doubt that, on a smaller scale, this company would not be above playing similar tricks.

I returned to the honeybee and asked them about its demise. The corporate giant Bayer whose product Gaucho, an insecticide sprayed on to sunflowers, had been accused of being responsible for the destruction of hundreds of thousands of honeybee colonies in France. Honeybees regularly gathered pollen in fields of sunflowers down here in the south. As I had learned from François a while back now, a bee has an innate sense of direction, a honeybee out gathering will always find her way back to her hive. However, sunflowers sprayed with Gaucho were fatal to bee colonies. The poison attacked their nervous systems and they lost their sense of direction, could not find their way back to their hives and died by the millions of fatigue in the process of trying to fly home.

Both men shrugged. 'The case was lost in court. Insufficient proof.'

Correct. A month or so before this meeting, the case had been lost in court. (Or won, if you were batting for Bayer.) The reason was that there had been insufficient evidence to prove that it was *exclusively* this product that had been causing the deaths of so many millions of bees in France. This result in court meant that the product could be sold again. It had been withdrawn from the market during the time of the legal proceedings, but it was now back out there and being used once more on the thousands of hectares of sunflower fields growing in the French countryside.

The beekeeper associations do not have the financial clout of heavyweights such as Bayer and there is also a strong case to argue that this pesticide was not the sole trigger for the damage being caused. And this is where the chemical companies have the upper hand. For the present, it was proving impossible to lay blame at any one door.

Possibly – more than possibly – it was the combination of products sprayed on to crops plus the accumulation of permissible quota and the build-up of residues in the soil from the previous seasons that were killing off the bees and other insects.

'Do you have dogs?' one of the men asked me, interrupting my argument.

'Two. Our third was poisoned by slug pellets sold for use on lettuces. Available to anyone and bought by a neighbour, hidden within food left out to kill wild boar and most unfortunately found by our little hunting dog.'

They ignored this.

'How do you protect the two you still have against ticks? I bet you dress them with collars to keep the bloodsuckers away?'

I did.

'Those collars are toxic to bees. Did you know that?'

I had not known.

'You see, we are living in a world where dangers are everywhere. Organic is not possible. It's a utopian dream. And pesticides are not as harmful as you would like to think. Besides, and most importantly, we have to feed the planet and that means mass crop production. We and our associates are doing the world a favour. Without us, the planet would starve.'

I decided to leave it. As I was gathering up my papers, the commercial one of the pair, Jean-Christophe Two, said to me, 'I want to be straight with you. There is one drawback to our organic product.'

'Really, what is that?'

'It doesn't work on small farms or isolated trees because it is an attractant. It attracts *all* the olive flies to it, so if you only have a few trees and your neighbour has many and he does not use this product, you'll end up with all the local flies in your grove. So, it wouldn't work for you anyway. Also, you need an agricultural licence . . .'

'Which we have . . .'

This surprised him.

'Our organic product has been created to be sprayed from planes overhead on farms with six or seven thousand trees.'

The image brought Agent Orange to mind again. It also recalled Andalucía, where the olive groves are far-reaching and where not a blade of grass, not a bird or bee exists within them. Where the planes fly low, dropping thousands and thousands of gallons of liquid poisons, contributing towards Andalucía's swift passage to a state of desertification, desert conditions. A terrain where, eventually, within twenty years if the soil specialists are to be believed, no trees will grow at all.

Anti-landscape.

'Sorry, we cannot help you solve your problem.'

'But it is very efficient, even if I say so myself, and I am responsible for this product,' countered his technical compatriot. 'If one single bee lands on a sprayed branch, wham, he is dead—'

'Excuse me, don't you mean fly, not bee?'

'Sorry?'

'You said *abeille*, bee.'

Both men laughed, just a tad uncomfortably. 'Yes, of course, I was referring to the olive fruit fly, *Bactrocera oleae*. Silly mistake.'

Yes, silly mistake.

I rose, thanked them for their time and returned to my car. I had not known back then, as I sat at the steering wheel deliberating, that over thirty of this company's products were on a European Blacklist of Dangerous Pesticides.

It was half past seven. The park was silent, eerily so. No lights on, no activity within the corridors of the chemical giant's technopole. The managers and officers, sales teams and technicians had driven off for the night to their gated and guarded villas and apartments, secreted somewhere within this perfumed landscape, a ghetto of its own, locked securely away, guarding its sulphurous secrets, which they, working on the inside, were party to and we on the exterior were not.

The following morning, after not such a restful night, I put in a call to the little *bio* offices in Nice in the hope that the lovely red-haired technician, Nadine, who had organised the outing with Vincent to the organic farm in February, could drop by and offer me some guidance, as she had promised she would.

'She's not here,' I was told by a young man. I requested that she call me back. While I was on the phone in my den, I spotted through the window, ascending from beyond the fig tree, the arrival of the Portuguese accompanied by an entire troop of dusky-skinned men, standing in the rear of the truck like a herd of refugees crossing a border. Were they Arabs or newly arrived from Portugal? They spoke no French. Although Michel had rejected the masons' advice to knock down and reconstruct the fourth stable wall, the misshapen one, demolition was the name of their game. Drills, dust, everywhere. Hair-line cracks, long, snaking fissures were appearing along the plasterwork and tiled floors of the interior of the main body of the house, caused by perforation vibrations.

Whenever I called to Francisco, who was back at work with red, weeping eyes, or to one of the Josés to take care, to pay attention, they would shrug and tell me not to worry.

'But I am worried,' I retorted.

'It's an old house and it won't last for ever. That door was coming away in any case.'

'No, it was not coming away! It was perfectly solid until you began trepanning the building!'

'Trepanning, Madame, what is that?' They walked away, heads shaking.

Nothing I said made any difference. Finally, I retreated inside and left them to their destruction, obliged to keep every shutter on the east side of the house closed. This was the area neighbouring my work space. I was at my desk in semi-darkness, an illuminated electric table lamp and candles burning, imprisoned within dust and noise.

Nadine eventually returned my call. She was leaving the organisation that very afternoon, off to pursue an entirely different career, but she would arrange for someone to contact me. She also informed me that we must register the farm as *bio en conversion* or there would be no support system, and this involved the completion of various forms, which she would send through to me.

Minutes later, a heavy file appeared on my computer, which, when I finally managed to open it, contained pages and pages of questions. Even the bureaucracy within the non-pesticide world was stringent.

I rang Nadine back directly to assure her that Michel would handle this within the next few days, but begged that in the meantime, please, could her replacement, whose name was Cécile, contact me as soon as possible.

Of course.

I heard nothing.

When I called again, Nadine had departed definitively and I learned that Cécile would not be taking up her post for a week or two. Soon, the all but invisible flies would be circulating the groves, preparing to infiltrate the fruits.

To add to my frustration, I suddenly found that the only phone line we possessed, apart from our mobiles, had gone dead. No internet! I went out to see if I could find the cause of the problem and found the wires hanging loose from one of the exterior walls. When I beckoned to *petit* José to find out what had happened, he told me it had been getting in the way and, as my husband had assured them that we didn't need that line, they'd cut it.

'But why would he say that? It's the only one we've got!' I barked.

'No problem, Madame.' The lines were spliced back together with red sticky tape. This gave me internet connection but communication from this landline was grizzly. I begged assistance from France Telecom and went outside to do some gardening.

In the evening, I printed out the copious pages of forms

and set them aside. I would not show them to Michel yet, but time was not on my side if I was to turn around his decision and Quashia's determination to forge ahead with the spraying. Aside from selling up, I had to find an alternative, a role model, and fast.

8

There is a jolly Provençal legend that I am rather fond of. It tells of a fisherman from the village of Correns who was out late one night crouching on the right bank of the River Argens, beneath a starlit sky, fishing for eels. There he sat gazing into the clear water at a magnificent reflection of the full moon shining directly down upon his rod. Suddenly, this villager's line began to twitch and move. He had a bite! Excited, he leapt to his feet, pulling and drawing. At that moment a cloud passed overhead, covering the moon, masking its reflection. The rod went still and the poor excited fool lifted it obediently out of the water. The cloud sailed off. There was the moon reflected against his hook. Convinced that he had fished up the gleaming globe of night, the *correnςois* went dashing off to tell his friends his great good news. 'I trapped the moon,' he shouted. 'Trapped the moon!'

And so was born the tale of *The Fisherman of the Moon* or, in Provençal, *Lou pesque lune*.

Twenty-first-century Correns, in the neighbouring *département* of the Var, does not permit fishing within the village itself though the River Argens still winds through its ancient stone heart and is home to many schools of freshwater brown trout. I had heard about this award-winning location on many occasions and had casually intended to pay a visit. Now, the moment was ripe. I needed guidance and, if I was fortunate enough to reach him, I believed there was one man there who might be able to advise me. A man who had made an extraordinary and quite incredible contribution to Provence. In his own fashion, he had reached out and drawn, if not the moon, then a corner of paradise to earth.

After having acceded to the role of mayor in 1995, Michael Latz persuaded the *correnςois* community of eight hundred to go green. Correns remained to this day, as far as I was aware, the only village in all of France that was operating entirely organically. It billed itself as *'le Premier village bio de France'*. Perhaps even more extraordinary was that Michael Latz was not from Provence. He was not even French, but German. So famous had he and the village he reigned over become that they had been visited by Ségolène Royal on her (unsuccessful) campaign trail to beat Sarkozy to the French presidency. The challenge for me would be to acquire an interview with him and at short notice.

Beyond the French windows, the dry late April morning was murmuring as I waited by the ringing phone. Eventually, it was answered by a young woman.

'Town hall,' she trilled.

'Yes, good morning,' I was a little uncertain how best to proceed. I gave my name and then dived directly in. 'I would like to have a word with the mayor, please, if that's possible.'

'Concerning?'

'His approach to *bio* farming.'

'Are you a journalist?'

'No.'

'When would you like to see him?'

'As soon as possible. I realise that his commitments must make my request a little difficult, but—'

'I'll give you his mobile. It's probably best if you call him and fix the appointment yourself.'

Quite taken aback by an offer of such direct and immediate communication, I dialled the number. The man in question was on answering machine. I left him a message and returned to my desk.

I was trying to make sense of the questionnaire sent through by Nadine: a richly structured environmental maze! Michel always dealt with the farm paperwork but he was not here and, in any case, he had decided temporarily against *bio* ... An almighty crash beyond the open doors caught my attention, quite startling me. It was fast followed by another. Swiftly on its tail came the sound of furniture being dragged, the dispersal of what could only have been broken pottery.

'Those blasted Portuguese,' I cursed silently, but they were not here. It was our dogs running amok. I rushed to the window and saw nothing, an empty deck. Nothing because a twelve-seater wooden table with eight chairs around it had disappeared from view. I threw on some shoes and scooted outside. Mayhem. Homer, who had been tied up due to the recent problems with the Rottweiler, was running for his life. A chain clipped to his collar, possibly three metres in length, was clunking along behind him, bumping and slapping up steps. Still attached to it was a substantial branch ripped from the cypress beneath which our Alsatian had been installed. He had torn himself loose! A swift glance showed the table was on its side, as were all its accompanying chairs. Terracotta pots were in smithereens. Hummocks of earth, shreds of wrecked geraniums and lavender peduncles, green stalks, savaged leaves everywhere. Paw prints autographed this mess.

'Homer,' I yelled. '*Homer, viens ici.*'

I had lost sight of him though now I could trace his trajectory by the paw prints, of which there were dozens and not all his. Lola! Lola was sleeping under another of the garden tables. The brouhaha had woken her but, lazily, she had not stirred, opening only one sleepy eye. I was confused until higher up the hill I spotted our black dog, lean and muscular, haring across the terraces towards the ruin on the Second Plot. Chain and branch still dragged along with him, damaging all herbage in its wake. Seconds later, I understood. On his heels was a black mammoth – the Rottweiler. In spite of Homer's branch and chain handicap, the heavier, stockier beast was unable to overtake him and remained an arm's length behind Homer, snapping, growling, at his haunches.

I yelled for our dog to come to heel one more time but he flagrantly disobeyed me. I picked up a pebble and flung it hard, intending to alarm rather than to wound. The Rottweiler looked my way, hesitated as though deciding which of us to attack, and then he spun about, changed direction altogether, hurtled up the land, galloping heavily, slipping off our plot into undergrowth towards the rear neighbour's newly constructed horse boxes, out of sight.

I returned to the furniture and began lifting the chairs and setting them upright. I was calling to Quashia, 'Please come and help me with this mess! Have you seen the workmen today?' He shook his head.

Homer was bounding down the hillside, triumphant in his victory. In the descending silence, I heard the neighbour's boyfriend, yelling, bellowing commands at his vanished black monster. I strode down the drive and found 'Johnny H' standing outside his gate. He was leaning on a pair of crutches. One leg in a plaster cast, clearly broken. I offered the whereabouts of his dog, or, rather, the direction the animal had taken. He scowled at me, and, immobile, yelled up the lane furiously then hobbled back into his own garden, slamming the gate behind him. I refrained from notating the damage wrought to our garden.

Returning to the house, ascending by the Italian staircase in this surprising heat, I found my hair covered with olive petals. The flowers on the olive trees were veils of creamy lace. We all three had agreed that during our years on the farm there had rarely been such a *floraison*. I was delighted and perplexed. We had pruned hard. I had encouraged it, hoping for small returns so that, if we were obliged to let the crop fall, the loss would be minor, but the exquisite sight of the trees was a bold denial of this fact. Nature was playing a game, promising an unprecedented harvest.

'There's no question we're letting this lot go,' crowed our gardener.

Without spray, how were we to protect these young olives? I could not deny that I was concerned, but if a choice was to be made, in my opinion it was to be our minuscule contribution to the earth and perhaps also to the resuscitation of the bee kingdom.

Inside, my phone was ringing. It was the mayor of Correns.

'When would you like to come?'

Sooner rather than later, if I wanted to enthuse Quashia and persuade Michel.

'Tomorrow?'

'I'm afraid that I have rather a busy day tomorrow. I could see you during my lunch hour.'

I grabbed the offer and we arranged a rendezvous outside *la mairie*, the town hall, for midday the following day.

'Meanwhile, have a talk with Madame M——. She can introduce you to everyone beforehand, help you understand how it all works here.'

Unfortunately, Mme M, the press officer for the village, who also returned my call directly, was already booked for the morrow and in any case she had a full-time job elsewhere.

She handled the village's publicity on a part-time basis, she explained, because she was enthused by its achievements. She emailed through a press pack, cautioning that if olives were my interest, Correns, Courrens in Provençal, might disappoint.

'I do hope you understand, Madame, that olive oil is more a sideline here. There are no *corrençois* who earn their income from the olive.'

White wine was the main produce and, second to the grape, came honey and goats' cheese. The area, I understood from her, also farmed cherries.

'There is not even a mill, *moulin*, in Correns but a neighbouring village, Le Val, operates one and that is where the *corrençois* who have olive trees go for their fruit pressings.'

'Is it organic?'

'Not necessarily. The region is renowned for its white wine, you know. You will be obliged to drive right by Le Val on your way into Correns from the motorway. Directly at the entrance to Le Val, where the road you need branches off to the left, you'll spot *l'Hôtel des vins*. It's a lovely building decorated with hand-painted frescoes on its exterior walls. It operates as an outlet, a *cave*, for the Correns wine cooperative. Attached to it is their olive mill, but it won't be open this time of year. There is no activity outside of the pressing season.'

'No little business selling olive-wood crafts?'

'Everything that is for sale can be found in the *Hôtel des vins*. Do pay it a visit before you drive on up to us. Sorry I can't meet you, but if you don't find everything you need, come back on Friday and we'll have coffee before I start work, about eight a.m.'

'That's very kind of you, but I hope it won't be necessary.'

*

The work seemed to advance in fits and starts. We would see enormous changes and then nothing. No one appeared for days on end. Today was one of those days when the Josés deposited their Arab skivvies and left them to get on with it. Men who seemed to do nothing but shunt mounds of sand from one corner of the semi-built site to the next. And then there were days, like today, when they reappeared at midday – the sacred hour of *déjeuner*. The second the hands of the clock in the local village touched noon, a siren sounded all across the hills and workmen downed tools no matter what they were engaged in. These Portuguese and their Arab helpers were no different. On days such as today when the work was dead, it was a ghost site until about quarter to twelve, then the lorries and trucks came roaring up the drive. Out would be unloaded baskets and bottles. A fire was lit, our barbecue and dustbins brought out for use, and lunch preparations were underway. I was beginning to feel as though we were running a courtyard restaurant. While the men caroused, I usually went down to see Quashia, knowing that all the imbibing was distressing him.

'I think it's time to cut the grass.'

By now, the herbage was brushing my lower calves and the heat of the sun was beating down against my face, creeping into corners, sliding itself into the nooks and crannies of my flesh, causing a mood of lethargy. Yet it was only late April. Insects – midgies we called them when I was a child, back when the grass was always as high as it was today, waving gently in an imperceptible breeze – were hovering in the still air. Chaffinches were chirruping. It might have been midsummer, a day becalmed, save for the absence of the cicadas. Along with everything else, the wild garlics were in full blossom, pale rose and white, and their sweet astringency perfumed the air.

'I would prefer to leave the growth just a little longer. It's not going to seed yet.' If I had failed so far to persuade the others to take the organic route, we could as a minimum follow Vincent's advice and create diversity in the groves. If we waited till every plant had gone to seed before strimming, we would be offering ourselves a far richer palette for the following years.

'I don't know why you won't let me cut all this back.'

'It's for the bees and I want the seeds to fall and create more flowers for their next season.'

'Bees! Are we having bees?'

'Yes, we will when I find someone to bring us hives and teach you how to be a beekeeper, Mr Q.' This pleased and quietened him for a moment or two.

'Well, at least let's dig up that red bush in front of the wood stock. It serves no purpose.'

Quashia had been much better mooded since Michel had told him that we would be farming the olives by the customary method.

'The pomegranate?'

Our Scarlet Woman, our sole pomegranate, with her striking frilly flowers, red as a tart's lipstick on a sailor's cap, she was also coming into flower up near the *hangar* where she had been replanted, alongside the soft fruits: the peaches, nectarines and deliciously sharp apricots. She had never given us any of her leather-skinned fruits packed with seeds and juice, a potent cocktail, and this was not the first time Quashia had nagged me to 'dig it up' because it was 'barren', but I refused to. Fruiter or not, she was a beauty to behold, and that's how it goes with Scarlet Women – they share their gifts with all who look upon and admire them. I could not expect her to be a bearer as well.

'No, I prefer to keep her.'

'But what on earth for? It serves no purpose.'

The pomegranate carried with it a trail of ancient history. It originated in Iran. Its fruits, packed with dozens of seeds secreted within little bags heavy with a tart, meady juice, were carried by caravan peoples thousands of miles across the deserts to provide liquid refreshment throughout the long weeks of arid journeying. Ancestors, quite possibly, forefathers of Berbers such as Quashia.

In any case, the poor tree had been dislodged once already. I reminded myself of the days when parties of olive experts tramped this farm, doling out advice, instructions, what could or could not be grown in between the olive trees in our groves; groves that were being designated as potential AOC oil producers. I believed now that much of that advice was unnecessary, if not downright questionable. Lifting out other trees, such as the pomegranate, reduced the fields to a single crop and that is detrimental to the longer term health of the soil. Removing that flame-flowered beauty from out of the grove because it was against the rules of the AOC bodies had been shortsighted. Diversity in the fields, as I had witnessed on the organic farm at St Jeannet, nourished the soil. The conventional measures were nutrient-destructive, eventually weakening the plants. In man's efforts to protect the weakened plants, to help them survive, to fruit, we feed them with chemical fertilisers, we annihilate the 'bugs' living in the fields, and so we are locked in an ever-descending spiral that is poisoning the environment.

'What are those men cooking now? Have you seen the bottles?'

'Where are the dogs?' I caught sight of Lola, prowling round the succulence grilling on the barbecue while the men, oblivious to her malevolent intent, continued drinking beneath the trees. I hurried off after her. 'Lola! Lola!' At the sound of my voice, she sloped away into the bushes. 'I'll be glad when all this work has been completed.'

'Carol, when you're shopping next, think of a couple of anti-tick collars. If you're insisting upon the long grass . . .'

'Actually, Mr Q . . .' but I decided against saying anything on the subject of the dangers to bees from dog collars.

While I was preoccupied with these details, a couple of France Telecom technicians had arrived at twenty minutes to twelve, bitching and begrudgingly. They clocked the emergency red tape, threatened me with a massive bill for destruction of the company's property – the wiring – then advised that if I got the job done on the black (either of these guys would be able to help me out for a reasonable cash fee on any Saturday) it would be considerably cheaper. If that was out of the question, they would sacrifice their lunch hour right now, handle the problem, log it in the book as all in order, be on their way and the matter would be forgotten.

So, what does a loss of lunch entail? I asked. A mighty outburst of shouting had broken out down near the parking area.

We agreed a cash settlement of seventy-five euros. I left them to it and scooted off to find out what the outcry was about.

Lola had filched every last chop and sausage the men had been preparing on the open fire in the yard. I found the three Josés, two young, rather bemused Arabs flanking them, running to and fro, knives, forks, frying pan in hands, charging around the lemon grove after Lola, whose slobbering jaws were jammed with meat.

This had been a theft waiting to happen.

From the fridge I pulled out steaks, two *faux filets* that had been earmarked for our supper. For a bunch of working men, I knew this was insufficient and I ran downstairs to find what other provisions I could come up with. Fish was unsuitable for the open fire. A chicken was all that I had to hand. Hastily, I lifted it from the spare fridge, grabbed a chopping knife and quartered it. Food on the plate, I carried it across the open verandah to the shaded spot where the masons had returned to their bottles of wine looking disgruntled, having resigned themselves to a liquid lunch.

I felt obliged also to offer a bottle of red wine, a rather good Bordeaux because there was little else, as compensation.

Our wine, consumed along with the other bottles they had brought with them that morning, was probably not the best idea. After lunch, the men were visibly inebriated. The smallest of the four tripped over a bucket of water and landed on his backside.

May had opened with a burst of searing temperatures. After months of intermittent rains, heavy, louring skies and downpours that had drenched the earth for weeks on end, summer, it seemed, was here. The following day, Thursday, I set off early, keen to spend a little time strolling about Correns village and its surrounding countryside before my noon appointment, relishing the peace away from the farm. I was really looking forward to meeting this man. In my eyes what he had achieved was a remarkable feat, and worth the outing just to hear his story, even if eventually he could not guide me because his expertise was in wine and not olives. I had looked him up on the internet the night before and as I drove the shortish distance along the A8, I considered what I had learned of this German who had captured the imagination and hearts of so many Provençal people, a race who, by nature, I had found, leaned towards xenophobia.

Michael Latz was born of German parents in the small landlocked African state of Burundi. Burundi, one of the ten poorest countries on earth, was annexed to Germany in 1899. During the First World War, the area was conquered by forces from the Belgian Congo. At the end of the war, after the defeat of Germany, the Treaty of Versailles divided up German East Africa. The British were handed the greater part but control of Burundi, along with Rwanda, was given to the Belgians. A Belgian suzerainty until 1924, the League of Nations then issued a formal mandate granting Belgium full control of these black nations, from thereon to be known as Ruanda-Urundi. Burundi's independence, along with that of Rwanda, was granted in 1962. A century of fighting, internal tensions, ethnic persecutions, its fight for independence both from Rwanda and Belgium, tribal differences, coups, a National Liberation Army provoking wars, followed by genocide, abuse of women, fleeing refugees – in short, the history of many post-colonial African countries – had brought Burundi to its present state of collapse resulting in a shattered economy.

By what route had a child of such a history ended up as mayor of Correns, I asked myself.

My journey was taking me seventy and some kilometres west of our farm. Once off the *autoroute*, direction Brignoles, the countryside I entered was nothing exceptional. I passed by the usual offerings of pink and beige modern villas, estates constructed swiftly to take advantage of the property boom in Provence. Eventually, I followed a sign to Le Val and began to climb a back road winding through natural oak woods. The heat was kicking in as I approached the *Hôtel des vins*, on the rue République, which displayed, as promised, a lovely exterior of pastel-shaded frescoes: bunches of grapes, pillars. There was a steady stream of customers, many of whom paused outside to shop at a stall selling locally produced fruit and vegetables. I might have been tempted myself had we not been growing almost the same selection at home. Within the *caveau*, there was a wide choice of wines, most of them reasonably priced. All were organic and all from Correns. There were also three different olive oils for sale. I was getting excited and went a little mad, filling up the entire boot of my car with rosé to take home to Michel. I tried to peer in through the barred windows of the adjacent mill – it looked derelict rather than closed for the season. It was impossible to see anything so I set off again. Now the route, still climbing, was lined with elegantly pruned olive trees. Their lightweight silvery boughs were dancing in the breeze, moving like busy fingers. I had reached a small plateau. Flat land now, save a distant valley or two populated with endemic cork and green oaks. The scenery was not remarkable, but it was wide open and had a calming effect. Its serenity seemed to take the edge off the blanket of heat. The dashboard clock in the car was registering thirty-four Celsius. It was early in the year for such temperatures.

I turned left at the sign and began my final slow crawl into the village. Correns had been settled within a fertile plain, today a basin of vineyards. In the distance, a semi-circle of bilberry-blue mountains, a range known as the Massif du Bessillon, and it sat within *La Provence verte*. There were signs at the roadside telling me so.

The greater extent of the village had grown up on the right shores of the River Argens, just as in the legend of the moonstruck fisherman. I passed by its calm waters after acres of vineyards, saw lovely curved stone bridges, old stone buildings and brightly painted houses. The long road in was promising. I turned left at a T-junction where a grocer's store on the right was decorated with a *trompe l'œil* that amused me. I drew up and stepped out to photograph it. An elderly woman in Provençal costume was sitting selling baskets of garlic and tomatoes while a younger man in a green apron stood in front of a green door smiling out on to a make-believe world that was benign, good-humoured. In reality, there was no grocer about to ask directions of. An arrow pointed me left to the town hall. I carried on past it, turned left again and drew into the free car park. From there, I returned to the *mairie*, a three-storey, bright yellow building. Here, in this narrow main street, the tall stone buildings kept the harsh heat at bay. They would also protect against the mistral winds that blew ferociously, slicing mercilessly through the veins of this part of central southern France. Walking or talking were young men, bare torsos, in shorts. Every one of them looked as though he was an organic gardener. Everywhere I walked there were flowers. This had also been one of Latz's initiatives: the seeding of medicinal and aromatic herbs and flower gardens. I strolled the length of the streets and discovered several fountains, alongside of which the aromatic plants were growing in recycled pieces of farming equipment. Favoured by landworkers, Renault 4s were parked outside the exteriors of houses with walls and shutters decorated in rich vibrant colours. A Provençal version of Painted Ladies! Fresh fish was being sold off the back of a large white van. A quartet of men were engrossed in conversation with the fish vendor. No women shopping; perhaps they were home preparing lunch? This was not a beautiful village – there are many down here that do answer to such a description – but there was an easygoing vibe about the place.

My mobile began to ring. It was the mayor.

'Where are you? I am running late. Go to the bar a few doors up from the town hall. I'll be there as soon as I can. It'll be around twelve thirty. Sorry about this.'

I assured him that it was not a problem.

My day was his, it belonged to Correns. My concerns for the farm were, if not drifting away, then weighing more lightly. Perhaps we should move here, I thought, and become viticulturists. I strolled over to the bar and ordered a glass of the famous Correns white wine. 'Is it organic?' I asked the bartender. He burst out laughing. 'Sure, it is. We don't serve anything else.'

His customers were men, all were farmers or land labourers, stopping by for a noontime glass of Ricard or wine, but, in spite of lives lived in the elements, they all had clear, smooth skin. They embraced one another as they arrived or left, called easily to others across the street, everyone spoke to everyone, which in a village of eight hundred was not impossible. A couple of women appeared and took their seats. They received the same hailing welcome. I sat listening to the voices, Provençal accents, some almost impossible to comprehend. There was little other sound. The sleeping cats were silent. Nothing, save for the wind whistling softly through the narrow streets and the rattle of leaves dropping – too early in the season – from the plane trees as they scuffed the ground.

Am I imagining this, I asked myself. This palpable *joie de vivre*. My wine arrived. It cost me one euro twenty, and was delicious. It was not as exceptional as a fine white burgundy but it was *très buvable* all the same. I noticed carved stone reliefs in the buildings, hidden interesting details. I picked up a leaflet lying on the table in front of me. It was the bar menu. It gave me a few lines about Correns' wine history. The esteemed local whites first gained their reputation when they were chosen by the monks of Montmajour Abbey, and consecrated for the celebration of the mass. This commune today was producing 50 per cent white wine while everywhere else in Provence the fraction was somewhere in the region of 5 per cent. I pondered this information. The UNESCO World Heritage Site of Montmajour was some distance from here, closer to Arles. In 1840, the writer Prosper Mérimée, who was also an inspector of ancient monuments, put the Abbey on the very first list of historical monuments to be preserved by the state.

My train of thought was broken by the arrival of a German car, an Audi. It pulled up in between the bar and town hall. I smiled silently. Surely this would be the mayor? Out stepped a tall, slender man, mid to late fifties, who glanced in my direction and strode purposefully towards me.

'Madame Drinkwater?'

I nodded. He veered off to shake hands with and embrace everyone. Every name was known. His presence brought smiles to faces.

'Let me take you to lunch,' he said, when all greetings were at an end. We set off in his car over one of the stone bridges and out of Correns, travelling a country lane for a couple of kilometres until we pulled into a pretty little restaurant on the left. Here, too, *Monsieur le maire* was well known. One couple on an otherwise empty patio were dining. They were strangers. Latz shook their hands, welcomed them to the neighbourhood and wished them *bon appétit*.

Over generous plates of salad topped with local goats' cheese toasted with honey – every morsel had been produced in the village and adhered to organic standards – I began to ask questions.

'How does a man, a German, born within a dark African land of heartbreaking chaos, end up as mayor of a small village sitting on the banks of the River Argens in inland Provence? A village that promotes itself as the flagship community for organic living in France?'

His answer was long, surprising, and might have been the outline for an epic film, a history of the twentieth century. His parents, both deceased, had been descendants of Polish Jews. Their lives had been that of survivors moving from one continent to the next. For a while the family had settled in Kenya, where Michael had his first experience of organic crops. The family had farmed coffee, quinine and the flower *Pyrèthre de Dalmatie*. He and his parents had lost everything, fled Africa, settled in Rome and then relocated again at the end of the Second World War to St Paul de Vence, down our way. Latz had been ten at the time. From there, they heard of an estate for sale in Correns. That *domaine* was where he spent his childhood. It remained to this day the family home. He had trained as an agricultural engineer in Paris and Belgium before returning home to run the family wineries after his father died from a stroke.

'He was too young, had suffered for too many years. Building fortunes, losing them, dragging his family from pillar to post.' I heard the loss, the regret in Latz's voice.

I returned us for a moment to the farm in Kenya. '*Pyrèthre*: that's the chrysanthemum that acts as a natural insecticide?'

He nodded.

'Do you use it here?'

Negative. The flowers were not selective. They destroyed all insects. In the Virunga National Park, where these chrysanthemums had spread like weeds, they were killing off plant life that sustained the natural habitat of the gorillas.

'So, what do you use?'

Vines, unlike olive trees, were not difficult to maintain as an organic crop.

Did they use the American traps that the chemical specialist had tried to sell me for their olives?

He shook his head. They were detrimental to bees and several apiarists were living close by. He was not against experimenting with new methods, he assured me, but what he was dead set against were the companies who were not only using our world crises for profit but stressing the calamities.

'Monsanto, for example. Are you one of the many voices against Monsanto?'

'Yes, but it is vital to remain open,' he said to me. 'If they came to me with an idea, I would listen, even today, even with their track record, but, alas, my experience is that such multinationals are only self-interested, and that is dangerous for our future.'

'How did you persuade all these farmers to follow you?'

'Well, I was their neighbour, remember. They knew and trusted me. It was remarkably simple, really.'

Soon after he had been elected, Michael Latz summoned

every denizen of the village and surrounding areas to a meeting to present his vision. The wine industry had been ailing here and he was offering them an opportunity to preserve their Provençal culture and reinvent their business. 'One tight-knit community, small wineries and a traditional way of life.'

Ninety-five per cent came on board directly and they have never looked back.

I told him that I had sensed a quite remarkable atmosphere while I had been waiting for him.

'Yes,' he said. 'There is a quietude here and a profound sense of joy because people know they have chosen their own destiny. We are not in the hands of anyone.' In the beginning, when he announced his plans publicly, he had received calls from adjacent villages, hamlets and estate farmers railing against him and the Chambre d'Agriculture had been vociferous in its objections. 'You cannot do this,' they said. 'You must follow the established paths.'

'"But we are doing it," I replied. They tried to stop us but we forged ahead and now they recognise and honour us.'

'What were some of the earliest changes you noticed?'

'No pesticides used in the fields brought us the return of plenty of insects, birds, small mammals and game. One of the first groups within the community who were happy with what we were doing was the hunting fraternity. Today, there is plenty of game in the vicinity and the fields are buzzing with flora and fauna.

'You would have noticed the River Argens running through our village? It is crystal-clear and teeming with fish. One of the reasons there is such an abundance of aquatic life is because of the quality of the water. From our commune, no pesticides or fertilisers are sprayed so there is nothing to end up in the soil or run off into the water. Most insecticides are lethal for aquatic creatures. We test our water once a month and it is pristine.

'We have a local *chevrier.* He arrived here to join our community in 2007. His herd of Rove goats pastures freely and controls the growth of the wild woodland areas that can be such a threat in the summer when the forest fires and the mistral winds cause conflagrations to spin out of control. They deliver us the finest of cheeses, which we have just tasted, and they help us temper the risks here in this harsh-weathered region.'

'And the future?'

'I, we, are committed to sustainable development in all areas. We've invested in a distillery and the commune is going to make and market its own organic essential oils.'

'From the plants and herbs I saw growing everywhere in the streets of Correns?'

He nodded, sipping his rosé while pouring us both another glass of Badoit. 'We will also dedicate certain fields to this. There is an increasing demand for essential oils and organic is very expensive to produce. This revenue will allow us to improve our social infrastructure. The village school serves organic food in its canteen and the students are taught the principles of sustainable development, of care-management for their earth. Those youngsters are the future and I believe that tomorrow begins with today. We, all *corrençois* inhabitants, including our children, are jointly constructing a way of life that is harmonious, independent and carbon-footprint light.'

I was not only impressed but humbled by this man's energy and vision. What was I deliberating about?

'Oh, but I am an impatient man. I had hoped the changes would come about more swiftly. As I said, I had no difficulty whatsoever in persuading my fellow *corrençois* to jump aboard the organic train, but I genuinely believed that once we were up and running other neighbouring villages would follow us. That was fourteen years ago. I see signs of small shifts further afield but it has certainly not been as dramatic or as swift as I had hoped.

'In the Old Testament it says you save one man, you save the world. My experience of life has led me to believe that this is not true. I am not sure man wants to be saved, wants a durable future that will halt the destruction of this planet. I am not sure that the majority of men give a damn. As I get older I realise that I am disappointed. But it does not stop me fighting on.'

'Any thoughts to leave me with?'

'Best not to interfere with plants and the cycles connected with them ... We are interfering with complex infrastructures that we have yet to fully comprehend and appreciate.'

'And for our olive farm?'

'Olive farming is certainly trickier than wine but my advice would be, *begin*, get on with it. Go organic and let the problems iron themselves out as you come up against them. I strongly believe that if you make a decision, a wise decision, life falls into step with you. The most pressing challenge is to make the jump. The rest will take care of itself.'

After we parted, I paid a visit to the Latz *domaine*, which he had suggested I take the time to do. It was beautiful. I crossed the river by one of the stone bridges and walked for a kilometre or a little more and there it was, nestling within crisply green acres and acres of vineyards, encircled by mountains. Nothing else in sight. I purchased a wine cooler at the little estate shop, walked about the small but elegant gardens and coveted the house. I had no phone signal to enable me to call Michel, to shout loudly, 'We are not going to spray. We are going *BIO*!' I had yet to convince him. We had jumped passionately into the purchase of the farm; it was time to make another uncushioned leap.

When I returned to the car and telephoned, I did not mention Correns, not yet. 'It's your birthday soon and we're going away for a few days, just you and me' was all I said, but I felt lighter of spirit and more clear-minded than I had in a while.

9

Michel's birthday was to be a special one. One of those when the calendar falls away and another decade slips inevitably into play. He did not want a party, had not been intending to celebrate it in any particular fashion. I suggested a family gathering, but he shook his head. 'Quietly,' he said, 'I want to cross over into that next decade quietly', and so I had decided to set up some arrangements myself. A brief excursion. Travelling for any length of time was not feasible with the works and challenges going on at the farm. A short trip, then, was the solution. An opportunity to talk. I chose a location I had discovered during my quest-travels and had longed to share with my man. Sicily – not too arduous a trek, doable within a long weekend from the South of France. Not a circuit of the entire island, of course, but a selected destination or two. It was also an island I hankered to return to, to spend time on, to continue discovering. I had even harboured thoughts, fantasies, of living there. Michel knew how the island had touched me, knew that the olive oil produced there was excellent because he had also enjoyed the bottles I had returned home with. When I suggested a birthday weekend in Sicily, he accepted without hesitation. It was what we needed, time together, to celebrate our blessings and to consider the future.

I booked a hire car and a hotel for the first night in Palermo. The rest I was leaving to the whims of the day.

We landed on the north of the island, west of the capital, into the Falcone e Borsellino *aeroporto*, named for Giovanni Falcone and Paolo Borsellino, two anti-Mafia judges who were assassinated in the early nineties. At the entrance to departures, a plaque bears their names with the inscription: '*L'orgoglio della Nuova Sicilia*', 'Pride of the new Sicily'.

I was doubtful whether this sentiment was shared by all on the island.

It was at this time of year precisely that I had made my first visit. Spring, when the horned-goat winds blow and wild flowers pepper the verdant hillsides, whispering of warmer, sweeter weathers. A brief spell to be cherished before the scorched heat beat into the earth and shrivelled the souls of both man and beast. The Mezzogiorno. It had, as I knew it would, drawn me back. It had been waiting for me. Or I had been waiting to return.

It was early afternoon. We had risen at half past four, changed planes in Milan and had missed out on breakfast. Our first stop: lunch. Michel at the wheel, we decided against all motorways, took a minor, forgotten coast road that wound its way round the lip of the littoral towards the capital. Little better than a rutted cart track, it was flanked by lusty, dusty vegetation, semi-abandoned, single-storey residences and overlooked by mountains hoarding secrets. So forgotten was it that the likelihood of a good restaurant was slim and, eventually, I suggested Mondello, where even at this latening hour – half past three – we would be sure to find a beachside shack or resto serving fish and salad.

I had passed through Mondello briefly last time and knew it to have once been a fashionable resort, poised between hills and sea winds, for the *bel mondo* of Palermo. Nudging the tip of a cape on the western extremity of Conca d'Oro, it was also neighbouring the city centre. Day trips, lunch outings or weekends away were all within striking distance and feasible for the *palermitani*, the inhabitants, citizens of Palermo. After quitting our sabulous lane, on the outskirts of the renowned resort, snarled up within our first volley of traffic, hooting, convening, lunging from all directions simultaneously, we beat a retreat to *la spiaggia*, the beach, past elegant Liberty-inspired early twentieth-century residences, creeping north along the *lungomare*, the curved esplanade. One of Mondello's highlights is its Art Nouveau beach huts, set on a raised wooden walkway out on the water and connected to the lido by a bridge. In the centre of these, a renowned upmarket restaurant, a 1913 gilded fantasy reminiscent of a film-set extravaganza. I doubted Michel would be tempted. Instead, beyond the mind-numbing hum of youths on scooters, north of the lido towards the heart of where the original fishing village had lain, we found a modest establishment, rickety, wooden, directly adjacent to the crystal-clear, green-eyed water's edge. While cutlery and paper tablecloth were laid up, we waited. At our feet, beyond a breakwater lapped the pebbled fringes of the teal-blue Tyrrhenian Sea. We were crooked within a crescent bay sheltered between Capo Gallo and, to the south, Monte Pellegrino, sitting at the gingered, whiskery edges of a vast natural harbour, Conca d'Oro. Strategically ideal, it explained why every conqueror from the Phoenicians and Carthaginians onwards had chosen this station as their Mediterranean heartland. Leaning against rusted iron railings, I was recounting fractioned episodes from the island's history to my husband at my elbow. He was taking photographs, listening, considering the information and the passion with which it was being transmitted.

'You love this place?'

'I do. Even in its ruined state, it is an "indestructible treasure", and I hope you'll soon see why. And yet, it isn't indestructible. Nowhere is. What would you say if we found ourselves a farm, a home in Italy, or in Sicily, more precisely?'

He was watching me, puzzled. 'Is this why you suggested this expedition?'

'Not at all, but I am unsettled at Appassionata. I haven't entirely returned, lacking an even keel, difficulties in accepting what's demanded.'

We ate that lunchtime, as we did for the rest of our all too brief stay, simply, local dishes. Sea- and land-based produce. I ordered the fried black-ink squid, *nero di seppia*, which is not a dish I would opt for at home, and Michel expressed his surprise at my choice. Before my first visit to this island, I might have felt squeamish about devouring a creature with three hearts and its excretions of jet-black ink. Chewing its rubbery torso soaked in juices struck me as an almost visceral act, rather like drinking blood, yet I had no problem with it here. This was an insignificant shift, but one born of my travels, of the Carol who had spent so long away from home.

'Upon this island, a tapestry of peoples, Asian, African, European, conquered, dwelt and left their imprints. It is a tri-cornered maelstrom, a mass of storms, conglomerations of winds, volcanoes, of ships being driven off course or wrecked and sunk. It is an island strangled by a dark and bloody brotherhood, who speak of honour above all else, kill for it, and where its starving peasant class have struggled to survive against every oppressor since time immemorial.'

'It reminds you of Ireland,' smiled my husband, who was gently alerting me to the fact that my lips and teeth had been blackened by the ink of the squid.

'Yes, I can see that historically it bears many similarities.'

Although no plan had been hatched, I realised as we sat talking, gazing out to sea, that I wanted my partner, this wonderful man whose life I had chosen to share and who had chosen to accompany me through mine, I wanted him to give space to whatever change of direction was pulling me.

I needed him to understand my doubts and confusions.

'An olive farm in Sicily: is this what you are asking me to consider? To quit France?'

A lack of clarity about the new direction I was seeking, only a certainty about what I felt obliged to reject, I wanted to break the seal over the upcoming days, precious moments alongside one another. Begin by describing the people I had met. The men in Sophia, the mayor of Correns; I wanted Michel and me to grope our way forward together.

I was attempting, with *pena* and *fatica*, rather unsuccessfully to read in Italian, Goethe's account in his *Italian Journey* of his trip to Sicily. After wintering in Rome, he travelled incognito to Sicily in 1787 and sojourned on the island between 2 April and 14 May. There was no reason for me to digest the writings of a German philosopher in Italian. It was only that I had chanced upon the book that morning at the airport in Milan and purchased it, hoping to use it as a multifold exercise. My Italian was hopelessly rusty – even during my recent excursions I had wrestled with it – and I longed to call it back within my grasp. Once upon a time, it had been my second language after English, but today it felt removed by centuries not decades, as distant to me as the voyages of Goethe himself.

'Sicily is the clue to Italy,' he wrote.

I had lived in Rome in my twenties and had believed back then that I would settle in Italy, but this island discovered by me for the first time during my recent voyages, with its earthy, secretive people, its lushly heady wines, its chiaroscuro way of life and caliginous history, somehow struck me as the real thing. Authentic.

'Italy without Sicily does not leave an impression on your soul.' Goethe, 1787.

Goethe also hoped to find in the Sicilian landscape the unthwarted Nature he had been seeking ever since his departure from Germany the previous year.

'Unthwarted Nature.' I knew what I believed had to be done, but I was groping, one hand in front of another, tugging at Ariadne's thread – should we stay, should we quit? – and I longed for the perspective and guidance of Michel.

At some point in the late afternoon – already time had left our table – we returned to the road. It was Friday. The working week was closing down as, replete and content, we reached the outskirts of Palermo. I had no idea where our hotel, a small boutique affair not of any renown, was situated. Debating the direction and caught up within the vortex of this city's unhinged road manners, suddenly, out of nowhere, an argument flared up between us. We shouted and fell silent.

Given its modest tariff, the establishment was unexpectedly central, just steps from the opera house and, mood still strained between us, we found it eventually. To my surprise, they even offered free parking for our hired Cinquecento. It was judged unsafe to leave any vehicle out in the streets of Palermo.

Our evening was given to walking, to discovery, to the pleasures of one another's company though I was aware that Michel was a little taken aback by our dispute of earlier. I was also shocked that a matter so trivial had driven us to an unseemly level of anger and raised voices. It was not our habit, had bruised us both and, I suspected, flagged a deeper disharmony.

Our wanderings led us to Vucciria, a filthy, fascinating, vibrant market and residential quarter where bands of squatters and down-and-outs were assembling in dingy piazzas, lounging, sipping beers, dragging on fags, hollering drunkenly. Aside from these lost, unkempt creatures, eyes as hollowed and shadowed as the finest of Caravaggio's portraits, there was barely a soul. The echoing silence in semi-lit streets was broken by the mewing, hissing of scrawny cats, the growls of beige, tipped-with-white mongrels, skeletal, tongues lolling, fleeing from the sites of ransacked trash as we strode past. Cars, scooters flashed by, nudging against our elbows as though fully intending to pulverise us. It was a stinking death trap.

All market activity had long since been wrapped up. For now it was the hour of *la passeggiata* in every other part of Italy. But not here. Down among these decaying lanes, where the houses bullied against one another, jostling for space, exhaling foul breath, trouble was afoot. We came upon two gangs of young men on bikes and scooters congregating in a square outside a Baroque church, smoking, leaning, grease-haired. Were they *picciotti* or just regular kids hanging out on a Friday night with bugger all to do? *Picciotto*, a 'young boy' or 'lad', in local and Neapolitan dialect, an admired figure of dash and daring, the next generation of males in the neighbourhood. The word's origins lay with the soldiers of Garibaldi's army who were involved in the liberation of Sicily, but it was also used to describe those who at a junior level had been recruited by the Mafia. This was not an unusual story. It was a picture I had witnessed in many Mediterranean countries where the disenfranchised were chosen, trained as 'soldiers', in whatever fight or cause was worrying away in the region. Al-Qaeda was at work in the Maghreb recruiting in the poverty-stricken rural communities. Here, it was city youths subsisting on pittance.

One block further along we reached a long, straight avenue where unlikely posses of uniformed police behind perspex body shields, armed with guns, truncheons and other weapons, were lining the narrow pavements. I asked an onlooker what was going on and he gave a most bizarre response: students from two universities, Genoa and Palermo, were converging to fight it out. Fight what out? I wanted to know. The stranger shrugged, dragged deep on his cigarette. Folk huddled in small groups; a level of expectancy reigned like that of the prelude to a corrida: clash of forces, blood sports. We passed through several of the blockades and hurried away, not that we had an itinerary or plan, simply an urgent desire to be clear of any rioting, which, given the level of police presence, promised to be brutal, murderous.

This was Sicily. It cannot be escaped. This was a city whose history was drawn in blood, of rioting and revenge.

It reminded me of a local snippet I had picked up somewhere. In 1647, after Palermo had been struck by a series of riots, the guilds were given responsibility for keeping law and order. These guildsmen, the *maestranze*, were allocated specific gates and points within the city's fortifications, each to provide a watchful eye. The fishermen and coppersmiths patrolled the streets by day and by night and had the right to pass beyond the city precincts, to bear arms and, if necessary, to hunt down, kill members of rioting factions. Precursors of the Mafia?

Our supper was an early one because we planned to be on the road soon after waking the following morning. We found an authentic Sicilian restaurant down a semi-darkened side street where the black-clad *mamma* sat with knobbled hands resting on swollen stomach. In a rattan chair near the doorway, she was nodding to, surveying, appraising the diners as they entered. It was her husband and two whippety-looking underlings with pencil-thin moustaches, possibly brothers, possibly sons of the proprietors, in shabby suits, who performed the role of waiters while father was the maître d'. *Mamma* did nothing beside eyeball the eaters and sit at table herself when a groaning meal was placed before her late in the evening. Judging by the décor, coloured lampshades, painted ceiling, it might have been a family sitting room, Sicilian-style, and perhaps had been in an earlier incarnation. The fare was simple but hearty. After my robust lunch of *nero di seppia*, I craved something light.

'No heavy sauces, please.'

'Signora, this is Sicily,' spoke our host. (How frequently

one was to hear those words.) 'Here we never defile the quality of our vegetables by masking their flavours beneath thick, unnecessary sauces.'

That established, we accepted *pappa*'s suggestion of lasagne of minced beef, fennel seeds and ricotta cheese accompanied by a tomato and onion salad, local olive oil (supermarket purchased, I was surprised to see), washed down with a bottle of black-red Nero d'Avola. From my seat, I had a view into the kitchen occupied by one single man, a lumbering colossus with an enormous paunch and chef's toque, flattened and misshapen. He manhandled the ingredients between fists as thick as steaks. After he had cooked the produce, he slapped it on to plates as though to be rid of it; done with this, done with that! All the while, he was muttering to himself, talking to his ingredients, coaxing the best out of those delicately fleshed offerings. Surrounded by flames, a series of stainless steel burners, mountains of fresh, gleaming vegetables, swags of herbs, slabs of meat hanging from hooks, great greasy canisters of olive oil from which he poured generously, he struck me as a solitary soul.

'I wonder they don't have an arrangement with a local olive farm to supply them directly.' It was a question I asked myself in every establishment we visited and I never found an answer.

'I wonder we shouted at one another like that,' I whispered eventually.

'Please don't always be so certain,' my husband replied and rubbed the back of my hand with his.

'Certain? I don't think so. I am still trying to find my way back into our life.'

Saturday. We rose and hit the road. At midnight, Michel's new decade would be rolling in. I was hoping we would reach Siracusa, but as this was a gift to my loved one I left the decision-making to him. Following the coast, avoiding as far as possible Bagheria, with its suburban overspill, ghastly concrete semi-completed blocks of flats funded by Mafia heroin money, we reached Cefalù mid these northern shores by late morning. Cefalù, a rock, a town, craggy, high above the harbour, gazing out to sea, bouldered fragments of a prehistoric city. Cefalù, conquered, reconquered, it told the history of this entire island: Greece, Carthage, Rome, Islam, followed by the Normans. Set back from the beach and harbour, winding up through the narrow alleys where the balconies were overhung with gloriously hued potted flowers, ascending eventually into Cefalù's vast cathedral square, a haven of cool, soft stone and quietude. Our days were to be punctuated by meals and this one was taken at an *osteria* in this Piazza del Duomo, where few others were dining and our view was the magnificent cathedral conceived by Roger of Sicily, the Norman king who, driven south by a storm, beached here.

Onwards, passing by high cols into citrus groves and agriculturally rich gardens set around decaying aristocratic villas, we penetrated the National Park of Madonie. Thirty-five thousand acres of nature with villages scattered on high or nestling in the lowlands. Here, a glimpse of Sicily *profondo*, forgotten, unchanged, possibly uncompromised. First came the bucolic idyll of farming land with pastures so green they might have been painted in acrylic where, at the roadsides everywhere, shooting skywards, fennel bushes with feathery fronds and mustard-yellow umbels; golden chalices shining in the sun. A tea made from the seeds of this *Foeniculum vulgare*, *finocchio* in Italian, is believed in the Mediterranean to be a galactagogue, a substance to induce the flow of breast milk. As a medicine, this plant, its seeds, its essential oil, serves many purposes. Native to this southern Med, used also as a base for a sweet aniseed liqueur, it dwarfed the asphodel growing at its side.

Clumpy yellow broom bushes, *Baptisia tinctoria*, with their beaded blossoms, the golden necklaces of gods, perfumed the scene. Every sense was catered for here. Herbs and plants to eat, smell, heal, touch, as well as to feast upon visually. I was drinking in the sights, guzzling them, taking no notes for this was not a quest. This was a birthday gift to the man at my side, yet I felt a certain sadness. Sadness that my questing was over? Sadness that we were not relocating here? Sadness that I could not settle back at home, could not accept the constraints? Sadness that our all too precious weekend had got off to a rocky start? I let the questions drift away unanswered and as I did so a memory slid back of the jungled growth at Appassionata when we first found her. Within it, yellow broom had been growing wild, tall as bolting trees. It scented the garden, wafting heavenwards, and I could not get close due to the overgrowth to pinpoint which of the plants was giving off such an ambrosial perfume. I wound down the window now, stuck my head out, let the wind blow my hair, my nose do the work, sniffing silently like our dogs at home when the olfactory system had been alerted. Yes, it was there, cruising gently on the air, playing its part in spring. Curiously, since we had rigorously maintained our groves, unlike so many other plants the broom had never reappeared. I asked Michel if he remembered those butter-golden shrubs within the jungled land, all those years ago, and he confirmed that of course he did.

Olives and vineyards followed.

Somewhere within this Madonie conservation area, where rivers ran at high altitudes, was a source, a rivulet springing up out of the rocky outcrops, known as Fosso Inferno, the trench of the underworld. It was believed that Odysseus visited during his long wanderings. I had no idea where it was located and there was no mention on our map. The Trench of the Underworld, a potent image.

In the middle of nowhere, where silence and birdsong reigned, where nothing but stunted, gnarled olive groves stretched, within which wild flowers grew, we came upon a low-lying, sprawling farm. Tucked away beyond tall, dark trees and rows of elegantly pruned olives, it was tinctured with warm tones, many shades of earthy brick were its multi-levelled terracotta roofs. It promised, from its middle-of-nowhere situation, an environment of peace, harmony, an idyll. Might it be this farm with its pale washed walls that I dreamed of in my own odyssean quest to reach 'home'? (Does travelling create an inner state of homelessness?) We attempted to find an ingress, a break somewhere in the metal fencing, but there was none. Eventually, at the foot of an endlessly winding dirt track, set back, a rusted iron gate with a hefty padlocked chain. Hanging from it, a sign informing callers that the property was owned (or managed?) by an olive institution whose headquarters were in Rome. An authentic body or a front for nefarious dealings? I had no idea. There was no caretaker about, no caretaker's lodge, nobody at all, and I could find no bell, no means of requesting the gates be opened that we might peep inside. For a national institution, its activities were very discreet.

I knew that more than seven thousand acres of land had been reclaimed from Mafia syndicates throughout Italy and these were slowly, very slowly, being reallocated to such bodies as Libera Terra who were honourably producing good wines, olive oils and citrus crops, organically, too, from the farms. Whether this holding had been part of such a donation, or was in any way connected to Libera Terra, there was no way of knowing. And nothing suggested it might be for sale.

On we journeyed, winding, climbing, descending. A cubed sienna-tinted palazzo here, shining beneath the sun, reigning over a hillside and acres of fields; elsewhere, another upland estate, ochre-walled, encased within sloping grasslands carpeted with green and yellow growth.

And late in the afternoon a hilltop town lost in backwash heights. Michel felt confident we'd find a bed but it lacked any hotel or *pensione*. Not so much as a bed and breakfast on offer even though the town was of a substantial size. Here we encountered wizened and bent Sicilian *nonne*, grandmothers, with ash-grey hair swept back into tight buns, thick stockings rolled to the knees, leaning on knobbled canes, climbing the steep cobbled lanes. Here, we found donkeys as transport as well as the modern-day chaos of cars and scooters, and a fountain with twenty-four spouts that overlooked green hills and valleys cloaked in spring colours. But no hotel. I was secretly delighted because I was still hoping to reach Siracusa before the day was out.

And so we did. First passing by mighty Etna or, in Sicilian dialect, *Muncibeddu*. The Mountain of Fire. In the falling evening light, she was more like a wild virgin goddess than a mountain, with her white orifice gaping heavenwards, seeping snow, exhaling iron-grey plumes of whirling smoke. A creature of volatile moods, exuding both heat and cold.

Finally, the outskirts of Siracusa. The same story here as elsewhere at the city's edges. Looming before us were aggregations of high-rises, sordid concrete blocks that made the eye sore to look upon them. Who, on the outskirts of such magnificent cities and towns, would construct these? Who would secure the necessary planning permission? On this island it was always the same answer.

Midnight in Siracusa's Piazza del Duomo, one of the most beautiful squares in the world. People everywhere. I had not expected such activity. It was Saturday night, but even so. On the table before us lay two glasses of bubbling prosecco and two magnificently complex ice creams woven, striped, striated with colours and flavours.

One of manifold gifts brought by the Arabs to this island was an early incarnation of ice cream, or, rather, sherbet. To the Arabs it was *sarbat* and then to the Italians it became *sorbetto*. Its base was a fruit syrup diluted with water. From Mount Etna ice was taken and the *sarbat* was transformed by partially freezing the diluted water to create a slush. This new delight became known as *granita*. By the eighteenth century, Sicilian ices and sorbets were renowned throughout Europe.

I raised my glass and wished my husband '*bon anniversaire*'.

'This is very special,' he grinned. 'You know I love you very much, don't you? You are the soulmate, the generous spirit I looked for, dreamed of finding, but it does not make you easy to live with.'

At our table, in a restaurant steps from the Aretusa Fountain, we were offered olives. 'Olives swelling sleek and dark' – Homer. Ours were green and dark and larger than cherry plums. We drank ruby-red wine and sat together after lashings of pasta, also brought to the island by Arabs, replete and deliciously satisfied, enjoying the early afternoon heat and the views and the passage of people.

Michel, who had been working at regular intervals in China, gently suggested that surely it was the Chinese who gave us pasta? I was delighted to be able to refute this. 'The Chinese had rice, but no other cereal crops.'

Pasta, the real substance, was created from a base of durum wheat, which, back when the Arabs were a force in the Mediterranean, was cultivated only in these southern Med climes. Spaghetti, vermicelli, macaroni: all were invented by the Arabs as they learned to work and utilise the durum paste. And because at that stage, eleventh century, the Arabs were still here in Sicily or on the point of being ousted, the Sicilians had mastered along with their colonisers the skill of pasta, the preservation of the basic wheat food.

Out on the silk-smooth water, a collection of white yachts. One, two – no, three – were approaching the shoreline. Each was tilted perilously close to the sea's surface, skimming, tipping, leaning – they are going over! – until, at the very last moment, each in turn rose like a pure white swan, flapping upright sails, taking flight, tacking gracefully; a ballet, a *coupé jeté* in the wind.

I was profoundly happy to be back in Siracusa and I hoped that the birthday man at my side was beginning to feel the pull of this place in spite of its abject poverty, its decaying beauty, its streets as slender and bendy as pipe cleaners. There we stayed until the sun was low in the sky and then we walked, going nowhere in particular, desirous of ambulation. Sneaking into crumbling courtyards, entering fine palaces, I was showing Michel discoveries of mine from the previous year, catching up on the changes, the 'improvements', some of which left me a little depressed, precipitating as they did the approach of mass tourism.

Vendesi. Everywhere, *vendesi*. 'See, there! For sale. And there, another.' A studio, disintegrating shutters with only flecks of paint remaining, a flaking-walled apartment, a barn, a ruin bound together in a symbiotic marriage of stone and wild, thorny weeds; a twisted fig tree thrusting up through its cracks and crevices, and, best of all, a partially restored house of soft pastel pink, a crouching-in-the-crook-of-a-curve abode, along the *lungomare* on the far side, the less discovered extreme of Ortygia. I scribbled down several of the numbers where it suggested that it might be a private sale. I had no desire to waste time talking to Sicilian middlemen. Michel caught my urgency.

'What has come over you?'

In the serenity of the candlelit night, I felt able to unburden, to shed the load I had been carrying, to allow the partner I loved to share my recent encounters, to see into the trough I found myself in. I talked of the anger I had felt as I drove home from Sophia, the inspiration I had felt after my meeting in Correns, that fleetingly I had resented Michel because he, by his decision to continue with pesticides, was not expressing such a vision.

'If we had been fortunate enough to have children together, you would not ask me to feed them what you know is detrimental to them. Losing out elsewhere, I want to preserve

Appassionata, to do the best for her. Appassionata is my child, our child, and the decisions that you and Quashia make sit heavily on me. I would prefer to up sticks and quit, to find another property where olives are not farmed, where the acres have been dedicated to other purposes, where our land management can be kinder to the earth or where I can simply walk away from all such responsibilities.'

'You are not trying to run away, are you, not you?' Michel nudged his shoulder close against mine.

'I am looking for a quiet place. Somewhere where the decisions will be less challenging.'

'Such a place is a dream. There is nothing to be said against dreams, but we are their weavers, we hone and direct them ourselves. To many Appassionata would be a gift, the Olive Farm would be their dream, as it has been ours. It still is and this is but a step along our path. Perhaps I am at fault for not accepting the depth of your commitment to the farm, and I can certainly reconsider my decision. Yes, we can go *bio*, but we will possibly lose our crops.'

'I am ready to pay that price until we find an alternative. We didn't buy the house to be farmers, it was an unexpected bonus, a blessing, but a blessing I believe we need to put back on the scales.'

'Then let us do so' were his carefully considered words.

On our evening circumnavigation of Ortygia, returning to the far side of the islet to where my pink house awaited its purchaser (it was not to be us) and beyond it to a forgotten, out-of-the-way pizzeria with a façade climbing in vine, we found ourselves caught up at the straggly hem of a costumed procession flanked by lines of Sicilian onlookers, churchgoers, believers.

'We can't get through,' murmured Michel. It was a dense congregation. We had no choice but to shuffle at the pace of the rest, haltingly protracted.

Far ahead, four or five hundred yards, within the bunched, crawling mass of religion, held high in the darkness, as though walking on air, but in reality transported on a pedestal lodged between two long wooden poles resting on the shoulders of broad-backed Sicilian men, rose a resplendently tacky golden statue. It had the height and the sturdy dimensions of any short Italian woman.

'What is happening?' I asked a black-haired female in the crowd.

'A homage to Santa Lucia.'

Santa Lucia, the patron saint of this city of Siracusa, the blind saint, the saint of light. I had mistakenly thought that her feast day had been celebrated in Siracusa the Sunday previous to this one.

'Oh, there are celebrations and services in her honour all year round.'

'Lucia,' I tell the man leading me, 'was a young noble woman from the fourth century who refused to marry her betrothed, a pagan, a choice foisted upon her by her mother.' It is said that the virgin girl distributed her dowry and jewellery to the poor. This act of defiance infuriated her rejected lover who denounced her as a Christian. The governor sentenced the girl to death. The soldiers, unable to arrest her, stabbed her to death, then cut her throat and gouged out her eyes. A gruesome finale for a poor creature who was standing up for her desire to marry the man of her choice, or not to marry at all if that was her preference. I was puzzled as to whether, in this land of blood-linked family honour and leather-jacketed machismo, these believers really understood the nature of the female they were celebrating.

Michel was growing impatient. Since I had known him he had never been comfortable in the face of formalised religious services or gatherings. Pushing forward, we reached the apex of the procession where the boldly disobedient Lucia towered over us in brazen glory and where, alongside us, a knot of men in several types of curious hats were chanting frenziedly, some holding candles aloft. Perplexed children watched on from the perfect vantage point of a parent's shoulders while we were jostled and shoved. Everyone was being jostled and shoved. Michel sought my hand, vigorously chiselling a path through the body of people, slipping us through the narrowest of apertures until we had overtaken the feverish pulse of this assembly. The procession turned at the next corner, making its way to the cathedral in the Piazza del Duomo. We paid those behind us no further attention and continued hand in hand along our way, intent upon our al fresco dinner.

The little *trattoria* of our choice had only one other couple eating on its terrace. Italians. It seemed remote and forgotten, but from where I was sitting I had a full view of my recently discovered house with its petal-pink walls, while, in the other direction, the sea.

'I don't want to leave,' I admitted, raising my glass of prosecco in yet another toast to Michel.

'Yes, these days, this too-brief trip, have been magical and I can understand why you love it here. Thank you for offering me a tiny taste of this stop along your journey.'

'Could you envisage living here?'

Michel placed his wine down on the table. 'I would like to return and spend more time discovering it with you, visiting the classical sites that we have had no opportunity for on this occasion . . .'

'But selling Appassionata and moving here . . .?'

'No, *chérie*, to sell up and move on, in my opinion, is not the answer.' There was no invitation for debate in his response. His thought was final.

Clear and cool was the air. Our chosen pizzas were taking a while and we sat listening to the passage of night out on the water, the beat of the sea's tongue against rock, beyond the battlements, with further glasses of prosecco and red wine to keep us fuelled while waiting for our food. The stars, jewels within the pitch of night, gleamed high above the ornate glass lamps. The pizzas, when they were eventually served, were delicious, but it was a blustery corner and I was feeling a little tipsy and confused. I wrapped my black pashmina more tightly about me, coiling it close against my flesh to protect myself.

I felt *triste*. The following morning we would be away from here, returning to the west side of the island, leaving this old world. The fact that we still had one full day before an evening flight did not lift my mood. A longing, both jumbled and bruised, was unsettling me.

Retracing our steps along the waterfront, the rolling rhythm of the sea dislodging the pebbles against the shore was like dice clattering. The game of chance. Beyond the city wall, an inky silver-blue unfurled before us. It was a dark, theatrical sight, imposing, rippling, dense and velvety as heavy curtains. In the sky above, a moon shone down, accurately centred – Archimedes, Greek mathematician, physicist, engineer and astronomer, born in this city in 287 BC, could not have calculated its placement with more precision – a full honeydew-melon moon, letting spill its juice, its luminosity, across the water in a sweeping, cylindrical shaft of light. It might have been the extravagant train from a wedding dress or a stairway, an ascending pathway to the stars. But the spill stopped short, beyond the city wall, a tantalising four or five metres out to sea.

Why does it not come ashore?

'Shall we vault this stone waterfront, sink our clothed bodies down into the freezing oil-slicked water and swim out to it, hitch a lift on to its shimmering, golden robe of light and let it draw us heavenwards?'

Glimpsing a path to paradise, such a sight for a birthday.

One tantalising step beyond reach.

★

Breakfast on the rooftop, surveying the city echoing with Monday morning hammering and restoration. A new port for liners, tourists, was being constructed with ugly grey breeze blocks right beneath us where, the previous year, I had strolled the quayside in between the tightly knit orange trees and the lapping waters of the bay. The entire area was to be dedicated to parking! Perhaps we had seen the best of it, I sighed.

We took the *autostrada* rather than our preferred secondary routes, to reach the airport side of Palermo sooner rather than later. Clear roads, good time; this left us with four and more hours to spare. A leisurely lunch and a discovery or two. Time for Trapani? Perhaps not. Castellammare del Golfo, then, which sits on the water within the province of Trapani. Why not? Here, a marina was under construction. We parked down at the water's edge and strolled along the key. Scopello was within spitting distance. It was where I had been heading the year before to meet the granddaughter of an American Mafia notable. Until, at the eleventh hour, she had backed out, refusing to talk; a change of heart, without explanation. This area, this coastline, had bred many figures who had found renown within the Mafia elite in New York City. A gang war there had been christened the Castellammarese War. But that was the past. Today, it was a port for tourism. Might the Mafia have fingers in the pie of this new emerging Sicilia?

There were three restaurants on this stretch of quayside, alongside the defunct *tonnare*, the tuna factory, but two were completely empty. We mistakenly assumed they were closed. The third was bustling. Taking this as a positive comment on the food, we followed the crowd. I spotted at another table two rather dubious-looking figures, both dining on mountains of *spaghetti di nero*. I asked them whether they would recommend it. One called the owner, puppet owner, I suspected, and ordered for me. 'Add a bit of this, some of that, a little extra . . .' and then the diner smiled in my direction and made a gesture. The famous Sicilian shrug. 'It'll be *molto buono, signora*, you'll see.'

Two small-time mobsters with a stake in the place lunching together? Michel told me I was being fanciful.

The remaining tables had soon filled up with groups of men, municipal types, heavies in shades, dark suits, brown shoes, Sicilian style. Was it fanciful to observe their relationship, their embraces with the owner of the joint? A carload of police arrived, pulled up directly outside, barring the view to the water and the semi-constructed marina. They were escorted to a table, served liquorice-thick espressos and then sent off with bags of hot food. Our plates arrived. Mine was sublime. The same white wine as those who had ordered on my behalf. Still, the two other restaurants on this slip remained silent and unoccupied, aside from one pair of pale-skinned tourists.

'Perhaps the other establishments haven't paid their protection money,' I joshed.

I wanted to visit the *tonnare*. Alas, it had been transformed into a hotel. Instead, we climbed up into the old town, to take photographs of the view and to amble about the narrow inclines. Up in the old quarter, an Arab feel to the place: flat roofs, cubed houses, market stalls of vegetables, looking out over the calm Med towards distant mountains, hazy in the hot afternoon.

Down a cobbled cul-de-sac, more an elongated square than a street, at the end of which lay the upper level of a castle now transformed into a museum, I raised my camera to shoot an elegant row of trees, acacias, I think, too heavily pruned to identify, with Michel centre stage walking among them. The terraced houses, each with its own wrought-iron balcony and slatted shutters, were shades of a sun-bleached rose or pink, like the morning sky at daybreak. Blossoms and creepers straddled them. Pots full of flowering geraniums on every doorstep. It was a natural, uncontrived southern

Mediterranean spectacle. I tilted my camera's lens to capture the opposite terraces and caught within the viewfinder the image of a huddled quartet at an open front door. I lowered my camera and sidled forward, to eavesdrop on the conversation. Two of the four were young men, twenties, dressed identically in white shirts, one short-sleeved, the other creased arms rolled to the elbows, and black trousers. Legs apart, they crowded an elderly couple who were pinned against their open front door. The old man, a wasting figure, mid-eighties, clad in lightweight grey outfit that resembled a prisoner's uniform, was attempting to edge forward, arguing weakly in a strangled, overworn voice. His bespectacled, robust wife, black crocheted shawl about her shoulders, was leaning against the frame, looking perplexed, distressed.

'You owe eight hundred euros,' claimed one of the two. Rent collectors, insurance men, racketeers? Surely this pair were members of the famed *picciotti*? Foot soldiers of the Mafia.

One of the thugs stood arms wrapped across his breast while the other drew out a paper, a file. 'See, here are the figures. Eight hundred euros.'

'But it's not accurate,' whimpered the grizzled old bloke while his wife muttered words I could not pick up on.

The second of the collectors was sent to their car, strutting, marching determinedly, to fetch – what? – further paperwork, trumped-up proof, a gun, an eviction order? The couple were getting visibly agitated. So was I. Michel, who had reached the entrance to the museum, was calling me from the top of a flight of steps, beckoning me to 'get going'. I circuited the scene, snapping fast, aware that, were I to be observed, my camera might be snatched or accidentally knocked from my hands. What did I hope to achieve with the recorded information? There was nothing I could do to assist this elderly couple, but my heart went out to them. I switched off my camera and hurried after my husband who was by now inside the museum.

There, on display within a series of glass cases, was the history of tuna fishing and its role in this community's heritage. I wondered about the frail old man outside. Had he spent his working life at sea, a skilled fisherman out on the boats in all weathers, hauling in those goliaths, playing his part in the bloody slaughter known as the *mattanza*? And those two young collectors of debts who, in spite of their different sleeve cuts, looked as though they were in uniform – why had they been harassing the octogenarian and his wife? Questions I would never know the answers to.

Time to leave, but leaving left an ache within me, a heart sore. Time to return to our intruder: just a fly, a minuscule fly.

Time to go home.

10

In order to put the farm on the organic road, we were obliged to declare ourselves *'bio en conversion'*. *Bio en conversion* described the transition, a process that takes three years, from conventional farming to organic. Once back from Sicily, we set up a meeting with Cécile, the new technician from Agrivert. Like Nadine, she was a young woman in her mid-twenties, dark-haired with creamy skin and alert, piercing eyes. She, too, was fired by a passion for this alternative way of life. We sat her beneath the magnolia tree, gave her a cool glass of mineral water and set to work. I had supposed that as a technician she would walk the groves and discuss the condition of the trees, but not at all. The purpose of this rendezvous was paperwork: a long series of questions and explanations, followed by more questions. The usual. What is the square metreage of the estate; how much is dedicated to olives; what farming methods were we practising; what products had been used here? Which mill did we frequent?

'You realise you will probably have to press your olives elsewhere. I will send you the list of those offering *bio* pressings. I am fairly certain that Gérard's is not among them.'

I was sad to hear that we would be obliged to make this change.

A droll fact was that, no matter where else in France we bought a plot or chose to live, once we had declared these parcels of olive groves organic, whatever other groves we owned, they were also obliged to be registered for the same ticketing. This, Cécile explained, was to guard against deception. A farmer might declare one small zone *bio* while farming other areas conventionally and then attempt to palm off the produce from all surfaces as pesticide-free.

An astute assessment of the wily Provençal mindset?

If we wished to grow non-organic apples, vegetables, wine or any other crop alongside our organic olives that was fine, which I found rather curious. Surely the pesticides from other produce fell on to the same land surface?

We would be subject to spot investigations. The experts would have the right to enter at a moment's notice to test the soil or any oil of ours kept on the premises. They might take cuttings from trees, if they saw fit, while a body known as the Committee of Certification (it had a communist ring to it, methought!) might also request an analysis.

All of these activities had to be paid for. By us. Not once, but annually. And, warned Cécile, they could be rigorous during the first three years while we were *en conversion*. The approximate cost for all this was somewhere in the region of five hundred euros a year. I recalled those farmers who had sat round the lunch table those months back. The poor fellow whose *cri de cœur* had been: every day I ask myself whether I shan't just chuck it all in. How many bottles of olive oil did one of those men have to sell to earn himself the price of the label that stated *production biologique*? Frankly, it struck me as steep and I said so.

'Oh, but you can reclaim it all,' encouraged Cécile. 'And you can also apply for an annual credit against your income tax, an allowance of twelve hundred euros.'

I doubted whether those hay-baley blokes had 1200 euros' surplus between them on their annual taxes.

'Well, why not simply cancel the charges?'

Both she and Michel answered in unison. 'It's a different department.'

French bureaucracy! It was ludicrous.

'Still, why must it be so prohibitively expensive?' I insisted. 'If we were earning our living from these olives, we would not be able to afford it. Surely it discourages farmers from making the shift?'

She shrugged. 'It's a young department and there's not much support for us yet.'

What about Sarko's Round Table and their promises to raise awareness, to assist farmers' comprehension of the principles of organic farming, to reduce pollution levels in surface and groundwater?

She laughed. 'A lot has been spoken about it. Little so far has filtered down into the day-to-day economics of running organic businesses. Only 2 per cent of French agricultural holdings are *bio*, but the good news is that 7.6 per cent of this region, PACA [Provence-Alpes-Côte d'Azur], has made the shift.'

'Why is the percentage so much higher down our way?'

'Converting olive farms is, as you know, challenging. There is still a great deal to be learned whereas it is relatively straightforward to manage vineyards organically so that accounts for the higher figures in this southern area.'

The products on offer to aid us in this shift were those I had already investigated. There had been no new developments in the ensuing months. As Cécile enumerated them, I rejected them. The good news was that *Psyttalia lounsburyi* was not a washout. On the five farms where baskets had been hung, the flies had reappeared at two. This gave INRA and the organic minds cause for optimism. The experiments were to be repeated. I offered our holding but it was impossible. The same five estates were to be used during the five years of experiments.

'So what is your approach to be?' she enquired. Apparently, it had to be declared.

I threw a glance at Michel. 'One of our first goals has to be the regeneration of the soil, to return it to a pesticide-free state at every level. This cannot be achieved during the first year, I know, but the process can be got underway. Otherwise, nothing,' I stated.

She scribbled my response on to her form.

I had been advised by Vincent to grow a wide selection of plants in the groves. 'We will follow the philosophy of diversity, recommended by your colleague from Avignon and, at the termination of every harvest, Michel will ensure that any residual fruits – forgotten olives, fallen ones and those that have remained on the branches unharvested – will be cleared away and burned. This will deny the flies their wintering abodes.'

'Good, good,' she cried, still scribbling.

'When we find a suitable opportunity and a beekeeper, we intend to reintroduce honeybees to the land. Otherwise, I would like to leave the earth to heal. I would prefer that the olive trees redevelop their own equilibrium.'

'And that will be it?'

I nodded. Michel confirmed it.

She left us with a word of warning: 'Your production is bound to diminish in the early stages. That is to be expected and it is for this reason that the government offers the twelve hundred euros' tax credit.' She also gave us a list with the addresses of four companies. These were experts. Our strict instructions were to contact one of them immediately, to carry out an analysis of our holding. Once they had seen the property, counted the trees, gauged our potential level of olive production, they would request signatures on a contract, to be filed at the Chambre d'Agriculture and also in Brussels.

This would be our official declaration.

Michel telephoned one of the four bureaux the very next day and within forty-eight hours we had received an estimate for their work and a one hundred and forty-one-page form to complete. A book of rules accompanied it, forty-two pages of small print!

'I'm glad it's you who deals with that side of the farm,' I joshed.

Cécile copied us in on the appraisal she had forwarded to the organic officialdoms stating that we were 'perfect material'. The experts came, the contract was signed: *Engagement pour la conversion.*

It was rather terrifying, but the first step had finally, finally, been taken. We would shortly be back on the olive body lists. From this day onwards, we would be classified as Olive Farm: *bio en conversion.*

I was nervous, a little daunted, for reasons I could not quite identify – honouring my commitment to the earth, perhaps? – but I was deeply satisfied that Michel had made this gesture alongside me. He also wrote all the letters to the olive bodies which had rejected us or crossed us off their lists. *Bio en conversion*, he informed them!

'We'll see,' he said, 'where this takes us.'

With pen and paper in hand, I paid a solitary visit to our oil supply. In a cool, dark corner of our 'summer kitchen' within the basement floor of the villa, I set to a little arithmetic. I wanted to calculate, just in case. Forty litres remained in Paris at our broken-down home outside the city, which was sufficient, just about, for our needs when we were there, for possibly the next twelve months. Here at the farm, decanted and settling in five *inox*, stainless steel containers: three hundred and fifty litres of recently pressed oil. Aside from those, there were forty-seven recycled wine bottles, each goldened by three-quarters of a litre. Plus there were two magnums once rosy with excellent local wine and today accommodating our juice. My sums showed me that we possessed 388.25 litres of oil. Oil, rich in polyphenols, which would assure its longevity and its fine quality. With careful management, allowing for the fact that Quashia was supplied, as were several members of our joint families – and I hated being stingy, so I calculated accordingly – we could live off our stock, if the worst came to the worst and we lost everything because Dacus attacked and we did not counter-attack, remaining oil-autonomous for approximately two years. This haul would buy us the luxury of time to research and find alternatives.

Late spring in all its fullness was busting forth. I could feel the earth swelling beneath my steps. Our winter had been a rainy one, wetter than most, and the dormant plants had gorged themselves on the gifts from heaven. Now they had woken up, harkening to the bugle call, and they were impatient to spread their seeds. Walking the land with the dogs fed me energy. It was crawling with life and brought on an unexpected sense of excitement. 'This is a new time,' I repeated to myself. 'A new way is dawning.' Even the farming magazine we had been subscribing to for a number of years had begun to run regular features on the advantages of conversion. 'Cherish your earth,' stated the magazine, 'and remember that an organic garden is rarely a sick one.'

I took it as wholehearted support for our new direction. In the earlier years, after I first signed up for this agricultural weekly, I would have been hard put to find any articles about alternative methods of farming.

As Latz said, make the decision and life falls into step, but one burning question remained: would Quashia fall into step or would he carry out his threats and quit the farm? We both agreed that it was essential to choose the optimum moment to break it to him.

As the days grew longer and lighter, Michel and I took to

rising at dawn. Breakfast – coffee sweetened with honey, fresh fruit, goats' milk yoghurt – was over by six. Then both to work, grateful for the brief lull before the onslaught. Now that the Portuguese were at work on the upper garage level, directly beyond our bedroom and my den, the dust was crawling into my sinuses and grit ear-marked the pages of my books. I was obliged to relocate files, computer, photographs and all the rest of my working materials elsewhere. Some mornings, I transported everything I needed across the garden to our pleasance, our Bedouin Bar, with its splendid view over the sea. Quite by accident, I had discovered that our internet connection was picking up even at that distance so I bagged it as my serene corner of connection. I could not leave equipment or books there overnight due to humidity or morning dew, but it was such a pleasure to work from there that I had no objection to transporting everything through the lengthening grasses. Inevitably, on the days when I had set up this haven, the men did not show.

When I was not writing, I was grubbing in the earth, planting, digging in sixty young lavender bushes, a mix of varieties I had found at the nursery, introducing them in among our principal bed of mature lavenders. I bought a box of watermelon seedlings and suggested to Q that we choose a corner and prick them out. I was looking for distractions, creating new interests, knowing that the moment was approaching when he must learn of our decision, for the olive flowers had departed and loaded branches of tiny fruit buds had appeared.

In the early evenings after the Portuguese had packed up and the noise had abated for a few short hours, Michel and I strolled the grounds with the dogs though we could barely spot them within the growth. Everything was bolting, be it shrubs, wild garlic, feral herbs. I could not identify some; I had never seen them before. Since we first cut back our newly purchased jungle, two decades earlier, the land had never been given such free rein and I was revelling in it, as I supposed the plants were, too. Perfumes, scents abounded; it was a veritable apothecary. Jasmine was climbing walls, orange blossom had broken out: a heady concoction. The apple trees, though, were showing signs of leaf curl or their own version of it. I paused to take stock of the damage. It was unlike the peach and apricot fungus that had turned the foliage a lovely autumn red. The apples were hosting small grey mites living on the leaves' underside. I had no idea whether it would damage the potential fruits, small as cotton buds. Painstakingly, I picked from the trees every leaf that had been infested and squashed it between my fingers. The sensation made me want to gag and left my fingers sticky, viscosy. After, I returned to the house and washed my hands assiduously.

'If this heat continues Dacus will soon be raring to go,' I called from the bathroom. I was also asking myself whether the longer grasses, the unruly garden, were encouraging predators. We were possibly susceptible to new invaders while the land built up its own complex webs of resistances. 'Michel, we have to speak to Quashia.'

'Yes, I'm planning to have a word with him in the morning.'

And then, overnight ... a wild boar invasion. The long stone wall to the rear of the lavenders where I had been digging had been buckled in three places.

'We cannot uproot and move them all now,' I said to both Michel and Quashia as we gazed down upon the disastrous doings revealed in daylight. This was not the season to be lifting out mature bushes and shifting them. It was too late. Several had begun to peduncle, a few were in spiked bud while others, the English variety known as *Lavandula angustifolia*, 'Hidcote Blue', were colouring at the tips, close to flowering.

'We have no choice.'

The wall was in collapse. Several of the big old stones had tumbled into the lavender bed. We were fortunate that the plants had not been flattened either by the heavy footfall of the blasted swine or the rocks themselves.

'It requires immediate attention' was Quashia's diagnosis.

The wall was one of the longest on the land. Thirty-one metres, measuring it by his great strides; thirty-eight by mine. It was not possible to repair selected sections because it was drystone and might disintegrate if taken apart. There was no knowing what amount of invisible subsidence had been caused by hefty hogs.

'There's a month's work right here. I'll need three tons of grey washed sand delivered this afternoon and, for starters, get me sixteen sacks of cement.' His curt demand. 'When are you going to get this estate electrified?'

I did not argue. If the wall was not attended to directly, the collapse would destroy the entire lavender bed next time.

I glanced at Michel who was helping Quashia heave fallen stones away from the plants. I knew he would not disclose our news now.

'I'll call Mark,' I said to the men.

'Yes, lock the boars out and keep the dogs in. No more worries that Homer will go seeking retribution against the Rottweiler and those wretched pigs, bane of our lives, will be barred.'

Mark had been working with us for a number of years. He was easygoing and honest and usually dropped by within a couple of hours of being telephoned, unusual down here. While the men cleared up, he walked the land with me, measured the fencing required and told me there and then that we were looking at a minimum of thirty thousand euros. Given the garage investment, the price was completely out of the question. I admitted as much to him over a cup of coffee and he advised against the expenditure. 'It does not always do the trick anyway. The beasts frequently penetrate whether the fences are electrified or not. I am receiving complaints all the time from foreigners who have bought expensive villas, contracted me to electrify the grounds, gone off after the summer and returned to find all the lawns churned up. Get yourself a gun.' His parting words. I was glad Quashia, who was always itching for me to supply him with arms, was not within earshot.

The next debate was where to replant the lavenders. They could not wait. The shift would need to happen promptly, which meant preparing the ground somewhere else without delay.

'I'll do that,' I offered, 'when we've agreed upon a suitable spot.'

In all, there were forty-seven bushes to transplant, all of which must have been close to a decade old, plus the new ones I'd planted in the days before and those still in their garden-centre pots.

'I think you should just burn the old fellas' was Quashia's thought. 'Their stems are long and woody. What good are they?'

I was determined to save them if I could. 'Lavenders are excellent fodder for bees.'

'But we don't have bees. You never did anything about getting us hives.'

I ignored this and suggested that, before Quashia began on the wall, we determine the new site. There was not a great deal of choice. These *Lavandula* needed a sunny spot, not too shaded within the groves.

The year before, with much of the work being achieved during my absence, Quashia had reconstructed yet another of the damaged stone walls over on the Second Plot. This one was possibly the longest we possessed and it had taken him, along with weekend labouring assistance from Michel, almost four months to complete.

All these old drystoners were repaired with concrete, which I was rather unhappy about, but the fact was that they were otherwise dislodged again and again by the invading hogs.

Our farm must have been an ancient pathway or *draille* – *drahle* in its original Occitan spelling – of theirs. The truth of the matter was the boars had foraged these hillsides long before we had arrived on the scene.

We had designated this massively extended wall, and the stoned garden that ran the length of it, for Michel's eternally talked of vines. It sat at the lower extreme of what had once been the estate's vineyard. The question was should we now use it for the homeless lavender? This was what – a hastily made decision – we settled upon. However, because it was a longer garden than the one that had been damaged by the boars, once I had replanted all the lavenders and installed the sixty smaller bushes, it looked half empty.

I set off for the nursery, for extra lavenders and culinary herbs to fill the gaps. Then, a most unexpected encounter on my way. Tidying up the lanes behind our hill, I spotted Jacques, our gone-underground swimming-pool man. I pulled over and called his name, fairly certain that he had spotted my car but was refusing to acknowledge me. I sat it out and eventually he laid down his rake and plodded over.

I could hear frogs in the nearby lake. Bumblebees, several kinds, were at work in the hedgerows.

'We've missed you,' I called as he approached cautiously. 'How have you been, how's your family?'

I was shocked as he drew near, when I saw his face, always so handsome. He looked grey, emaciated. My guess was that he, his wife or daughter had been struck by illness, but I was quite off the mark. Jacques had been subject to 'an investigation'. He would not clarify. His company was 'in deep trouble'. I assumed that this work, cleaning the streets, was a compulsory community service he was undertaking.

'Come and see us. Will you work for us again?'

He shook his head. Those remarkable eyes looked hunted. 'I wouldn't dare,' he muttered.

On sale at the garden centre were gorgeous examples of a

French lavender, *Lavandula heterophylla*, or sweet lavender, a very robust plant. Another Frenchie, *Lavandula dentate*, 'Candicans', I chose for its almost silvery, creamy, light green foliage. Its blossoms were softly tinted, a washed-out blue. It had a slightly ghostly quality about it (like Jacques), and would complement the louder purples. I was quite astounded to discover the range of varieties. As with the bees – the honey and bumble were but two – I was equally ignorant of the *Lavandulas*. French or English had been the extent of my expertise and now I was stumbling upon choices from a list of more than two hundred varieties. The Lavender Route. It was as thrilling as learning a new language. But how shocked I had been by Jacques. In the days when he attended to our pool, he had become employed by a wealthy Englishman. On the black, of course. The Brit was retired from the secret service, according to Jacques, and was very liberal with funds. On the strength of this generosity, Jacques had started to learn English and was building himself a house in the hills near Gérard's mill. I wondered whether this had been the undoing of our lost friend.

And what of Luke? Might I one day bump into him and uncover a similar moral tale? Might he be in some rank African prison charged with smuggling, or had he been pulled over, 'nabbed', for driving a vehicle fuelled by vegetable oil while on a public thoroughfare and then found to be transporting several thousand illegal exotics?

The underbelly of the Côte d'Azur!

For the past few days, Francisco had been working our site alone. At first, his companions had roared up the hill around eleven, unpacked lunch, set the fire going and prepared the barbecue. The foursome – their Arab assistants were no longer employed here – ate together, *comme d'habitude*, in the shade, drinking copiously, of course, then siesta'd beneath the hanging boughs of one or other of the great olives. Four of them, heads resting against a monumental trunk, stretched out on the grass, snoring. Quite a sight. Once awake, a swift coffee downed, the three Josés departed, hell-bent on another job, leaving Francisco alone while the clouds of smoke rose out of the dying embers and blackened an ever-increasing circle of grass.

But, of late, Francisco had become a solitary diner.

'My companions are on another contract,' he told Quashia.

A state of peace had been established between these two old-timers (temporary, it proved to be), which was heartening to observe. Each morning when Francisco stepped out of his white van, he and Q would hail one another with hearty greetings before embracing in tight bear hugs. Together, they strolled about the site, arm in arm. Francisco with operatic gestures explained to his Algerian comrade his plans for the finer details, the finishing touches of the work and Quashia would nod his assent, smiling, laughing.

So you approve of what he's doing? I asked of Q, with a teasing wink.

'He's not like his companions,' our man said. 'He has a finer perception of the world.'

Francisco undoubtedly fancied himself as the artist of the team. He wore his hair long, smoked endless cheroots, complained about his mistresses and ex-wives whereas his pals were down-to-earth family men, earning a living, bringing up children.

Now that the *gros œuvre*, major works, had, according to the men, been more or less completed (which they had not), Francisco was engaged in the plastering work, the definition of corners, of pillars, wooden beams. He worked diligently, but we were now close to the end of our second month and the damp course had yet to be laid. There remained several important steps to be accomplished and I was puzzled as to why the men were not proceeding with them. Once or twice I offered Francisco a coffee, which he always refused, and enquired after the construction's completion. 'When will the others be back? The builders' yard is nagging me about the delivery of the terracotta tiles.'

'*Bientôt, Madame, bientôt*.' Soon. He was guarded, never more specific.

It was early afternoon. I had been working inside, out of the early June heat, and had gone out to check on the dogs. Quashia was complaining that the longer grasses had encouraged ticks. Every year until this one, at the outset of summer, I had bought collars for the animals – never sprays – but since my meeting with the agrochemical experts I had avoided this. Even though I had eventually explained my reasons to Quashia, he judged me cruel.

'We will have to be vigilant and check them every morning and evening,' I insisted, which was what we were doing.

'You think of the damn grass before your own beasts,' he had huffed at me.

Where Quashia was now, I had no idea. Perhaps resting after his lunch, unwilling to continue the wall demolition until later, when the heat had eased. Suddenly, I noticed water everywhere, running in streams down the drive. I spotted Quashia, hat in his hand, scratching his head.

'Has a pipe burst?'

'It must be him, drinking two bottles of wine at lunchtime. He cannot work, he's not capable.'

I was rather surprised by this, given their newly bonded friendship. 'Where is he?'

Francisco's van was visible but there was no sight of him.

'Drunk, sleeping it off somewhere,' growled Quashia. 'They've been here for months. We need that garage finished. I have to get on with preparations for the olive spraying. They're holding me up.'

My heart went cold at these words. We had still not come clean.

'Let's get this water turned off.'

We traced the leak to a small tap up beyond the works where a hosepipe was being used to soak the cement mix. It had been left open. We called for Francisco, but he was nowhere to be found.

'He ate alone and drank the wine that had been put aside by his comrades. A full bottle and the remains of their quota, too.'

'Nonsense, Mr Q. A glass with lunch does not mean he's drunk.'

We found Francisco sleeping in the back of his van, doors wide open, conked out. Strong coffee was required. I hurried to the kitchen to get it underway, but the taps coughed empty. Our basin – its maximum capacity was one thousand litres – had trickled its remaining contents down the drive. Quashia strode into the valley to switch on the waterhouse pump. I could not bear to contemplate the cost of such waste. When I finally delivered black coffee to Francisco, I asked him when the men would be returning to finish off the roof. His reply was: 'What do I know? I quit the company. This is my last assignment.'

I begged Michel to contact the Josés and find out what was going on, when the works would be terminated. 'And it is time to talk to Quashia.'

Michel nodded. 'Yes, he's entitled to know.'

Early the following morning, I carried out a cup of freshly brewed coffee to our man. Two magpies were screeching, squabbling over unripened fruit in the fig tree. He was down in the old groves, demolishing the remaining quarter of wall, preparing the whole site for reconstruction. It was time. I intended to explain clearly why we had made our decision. I wanted him, perhaps more than anyone else, to be with us on this. We needed his support.

He took the coffee with a wide-open grin.

'*Excusez-moi,*' he said as he always did when I brought

something for him, as though 'forgive me, for the trouble this has caused you'. A gentle consideration on his part, but all such empathy disappeared instantly from his features as I began to detail what this transition involved and why we were making it.

'But what about the summer spraying against the flies?' he demanded.

'No,' I replied softly. 'We will not be doing that any more.'

'But what about the spraying machine you bought me? What's to happen to that? And I thought I was to be laying paths up the hillside with Michel so that we can access the young groves at the summit ... To prune and spray, and harvest more easily. We've been planning it for more than four years.'

'Well, the trees will still need to be accessed. We will still be pruning and caring for them. In fact, now more than ever, Mr Quashia. The health of the trees will be paramount. The healthier they are, like any one of us, the greater resources they have to withstand illness, fungus attacks.'

'But not the olive fly! You know as well as I do that no amount of trimming or pruning is going to rid them of that little blighter. And if you think it will, you are being ridiculous.'

'Please, Mr Quashia, please don't get upset. There are new methods being developed, even as we speak, to combat Dacus.'

'Dacus?'

'Sorry, the *mouche d'olive*.'

'You mean like that fellow you brought here who promised to inundate the place with African flies? Carol, I have been farming since before you were born. Algeria is one of the most ancient olive producers in the Mediterranean. I know what I am talking about. You need to kill off these insects with poison, and that's that.' He downed his coffee in one shot, handed me the cup and returned to his digging. He was not to be converted. Certainly not by me. My convictions meant nothing to him. Worse, he judged me a dilettante, as René had done.

I suggested to Michel that he go to him. 'A man-to-man conversation about the challenges on the land. Please try to persuade him to work with us on this new programme even if as yet he does not agree with us.'

Michel walked down to his cottage later in the afternoon. His television blaring out in Arabic, a lamb stew boiling on the stove, they sat together and drank a glass of lemonade and Michel tried again.

'Ideas, just ideas. The farm demands action,' he mumbled.

Michel patiently explained why we needed to change our methods. 'Carol is concerned for you, too, Quashia. Your health. It is one of the reasons she is so dead set against spraying. These products are dangerous for everybody's health.'

But Quashia was not convinced. 'I am not a young man,' he confided to his employer and friend. Michel was a man Quashia valued as his brother, better educated, more able to work the French system which favoured Europeans, not the immigrants, the Maghrebians in particular, and he had been deeply grateful on several occasions over the years when either Michel or I had been at his side to defend his dignity against the officialdom that can be so dismissive of its foreign workforce. This, too, had bonded us. We were not divided by class or colour. We were, all of us, committed to Appassionata, its beauty, produce, and we all in our very different fashions loved the Mediterranean and its ways of life. But it appeared that we had reached an impasse.

'I have seen fads come and go and that is all this is. I do not want to waste my time pruning trees, repairing walls, whatever it might be, just to see them destroyed by boars or olive fly. I love olive trees. I gain great pleasure out of standing in the sun watching the branches swing. Well, we all do, I know that, but I grew up in a land with olive trees, an olive culture. Neither of you two did. I was weaned on olive oil!'

'We don't doubt that, Larbi.' Michel frequently addressed him by his given name.

'I know what I am talking about, Michel, and yet Carol will listen to the advice of any passer-by, but she will not listen to me. Hear me, please, you are an educated, intelligent man, you will understand what I am saying. We need to spray those trees or we will have no crops and, if we have no crops, you don't need me. I am better off spending these last years of my life in Algeria, at home with my family. There is nothing for me here. I am a simple man, too old to change, too old for this nonsense. I love Carol as though she were my own, you both know that, but, excuse me, Michel, I fear she has lost her way, her ability to make productive decisions. She disappears for months on end, lives in the clouds believing in utopias that don't exist and never will. If you want to allow this, then that is your choice, *mon chef*, and I must respect that, but you are the head of the household. Take charge, tell her firmly, "No, Carol, I am the man. I know best", and let us get on with the business of farming olives. Otherwise I might as well pack up and go back to my son's shop and end my years peacefully. There is nothing left for me here.'

It was probably the longest speech Quashia had ever made, certainly to either one of us. When Michel recounted it later, he said that our loyal assistant had spoken without anger but with certainty.

'We are losing him' was the summary of the hour in the cottage. 'He is resolved. It will be difficult for us all.'

'Is there anything we can do to change his mind?' I begged.

'You mean aside from returning to our old ways of farming, because that is what it will probably take.'

I shook my head. 'I cannot,' I whispered. 'I love him dearly, but I cannot.'

*

I tossed and turned most of that night. How to retain our man yet honour our decision? Quashia was not a person to quit. To leave us might be as painful for him as it certainly was going to be for us, and I suspected that he needed the money he earned from the farm. He collected two reasonable pensions, but he was the paterfamilias back in Algeria and the only full-time earner. One son of his had died a few years back in a tragic accident and his widow was bringing up four if not five children alone. Quashia was their breadwinner. On top of that, his youngest son, now married with a child or even two, was not the most responsible by all accounts. He ventured into situations without counting the cost. One moment he wanted a farm in the hills with sheep and goats. Quashia had purchased the stock for him and helped with the construction of the farmhouse, and, soon after, the offspring was investing, with his father's unearned money, in an *épicerie*, a grocery store. I knew those commitments weighed heavily on our friend and I did not want to be the one to deprive him of the wherewithal to feed his family, most especially the partially orphaned grandchildren. Women in Algeria, particularly widows with a gaggle of youngsters, were rarely given a second chance. Quashia's daughter-in-law, who had married the son before she was twenty, had never worked, never earned a living, never been trained for anything other than housework. Her roles had been cast: wife, mother and now widow.

A parting of the ways was not what Quashia desired any more than we did. Beyond our bedroom, beyond the semi-completed terrace, an early thrush was singing in the Judas tree. Aside from all of that, Quashia was my friend. Our friend, an ally to us both but perhaps more particularly to me. Losing him was tantamount to losing a beloved relative.

On the other hand, I had to be true to my own heart.

The following morning, Michel was on the dawn flight to Paris. After I returned from the airport, Quashia plodded up the drive, eyes sunk to the ground, and I knew that his decision had been made.

'I won't have time to finish the wall, but I'll cut the land back, then I'll be on my way. I'm intending to take the train on Friday evening.'

I nodded. We might have waited another week, perhaps two, before felling the grasses, but I accepted his decision. Vegetation maintenance was a legal requirement in our corner of the Midi. It reduced fire risk and Michel and I knew all too well the horrors of *un incendie*, a wild fire on the rampage. *La lutte contre l'incendie*, the fight against fire, was the responsibility of every citizen. We had left it to the very last moment, but now the job would be done.

It was Quashia's final chore, to pull out the strimming machines and fell the wild flowers, the herbs, the 'weeds'. Everything, more or less, had broken into seed and the seeds were departing in readiness for future springs. It took him several days and during that time we worked alongside one another in an amicable fashion, talking intermittently, and, when the grounds had been flattened, Q cleaned up the machines and returned them to the *hangar* because the garage was still not finished – Francisco was occasionally present but there was no sign of his former comrades. It had begun to rain. The first drops in a couple of months. Soft rain. Rain to swim beneath, rain to suckle the soil, rain to breathe the bittersweet scent of flayed green earth.

I knew he felt that I had let him down, betrayed our working ethic, and I forced myself not to doubt the choices we were making because this transition – this period that we were living through now – was proving to be far more painful than I had anticipated.

Before our Berber friend set off to pack his bags, I offered to drive him to the station in Cannes. From there to Marseille and then a flight to Constantine. I knew this might be the very last journey we ever took together and my heart was breaking over the loss of him.

His leave-taking was a sad and unsettling moment. His valedictory words: 'I'll get these wretched teeth out, see about dentures. Should you ever have a healthy harvest again, call me, tell me. Now I'm in retirement, I'm going to get myself some bees.'

I had no idea how we were going to manage without him, or if we would ever set eyes on him again. I put my arms around him and hugged him as though he were my father.

'Thank you for everything,' I whispered. '*Je vous aime.*'

He nodded and set off with his carrier bags, without a backward glance.

II

Francisco was packing up for the day as I returned from the station. He waved me over. I was not really in the mood for conversation.

'Your Arab told me he'd quit. Said he was unhappy about the choices you were making.'

It rather hurt me at this vulnerable moment to learn that Quashia had been talking the matter over with Francisco, had criticised me, even though I was aware how he had felt.

Francisco wanted to encourage me in our agricultural decision, he said, and cited his grandfather in Portugal. 'There were a few flies, but my grandfather paid little attention to them, never sprayed. He harvested and pressed the drupes without concern. The oil,' as Francisco recalled, 'was excellent.'

The old man's principal activity had been *pastèque*, watermelon. This was the sacred fruit, the red-fleshed fruit, sweet and juicy, which the grandchildren were forbidden to go near. Acres and acres of expanding green bubbles on the undulating landscape. When they were ripe, Francisco and his pals used to steal into the fields and dribble the melons gently like footballs across the sloping land to the hedgerows where they hid them until after the *récolte* had been completed and the old folk had disappeared with the bulk, transporting it like a mighty green balloon rising sky-high. Donkey and cart, to a market town twenty kilometres' distance from their nearest village. Then, arms waving victoriously, the yelling children hurtled back to the bushes and claimed their stash, splitting the cumbersome fruits open with bread knives filched from the kitchens and gorged themselves until their mouths and faces bled with the candy-pink flesh. The big black seeds they dried in the sun and used as counters for their games.

'We were very, very poor. Portugal, like Spain, had been crushed beneath a dictatorship.'

Had I known, this decorator asked me, that the watermelon had originally been cultivated in Egypt along the banks of the Nile? I had not.

'A wise choice you are making,' he smiled.

I thanked him for his story, bid him goodnight and went inside.

Still, Quashia had gone. There was an emptiness about the place. His laughter, his temperaments.

And how the workload increased! As the heat intensified, the watering commitments escalated. I began at daybreak and worked for two to three hours. I took delight in each patch, in the vine-like *pastèque* and its swift transit across the soil. I counted eight green balls, some the size of magnified eggs, others as large as pommels and the largest, rugby balls. They reminded me of our desert friend, as did the transplanted avocado tree, pushing up like a buster. Having discarded its yellowed leaves, it was producing growth shoots of a tender amber. The hand of Quashia was everywhere while mine were growing callused, my nails broken and seamed with soil, and no matter how thoroughly I scrubbed, the traces of earth never disappeared, but I didn't mind. In fact, I perceived it as a badge of honour, an accomplicity between myself and the land. I could use gloves if I chose. Indeed, I sported a sturdy leather pair (one of those Quashia had hidden from me!), but I always ended up chucking them in the wheelbarrow. They got in the way. I preferred clawing with my fingers down beneath the crumbly dry clods when I was planting, turning the soil over, feeling its textures, listening to it. Any weeding that took place, which was precious little these days, was achieved by hand, my hand, and I was far more relaxed about it, seeing little as opposition.

Every evening, after hours of labour (the irrigation of the junior olives at the apex of the hill took Michel three days every three weeks), Michel and I relaxed on the terrace side by side, enjoying a well-earned glass of chilled wine. While the frogs down at the stream in the valley crooned, we attended sunset and it was usually a spectacular display. Some evenings a pair of buzzards, 'our buzzards', circled in the sky overhead. I liked their call, the clean, distant whistle they emitted. Our two pairs of eagles were once more in evidence. I was still hunting for their nests but I knew they were not far off, breeding within our pine forest. At night the tawny hooted from the big old pine beyond our bedroom. It was second-brood time. Another owl had also set up home on our hillside. His call sounded as though he had a penny whistle caught in his throat and made us giggle. So much to relish, but the farm needed managing and, when Michel was away, I was hard put to achieve everything. We required another pair of hands. Occasionally, we employed casual labour, but I kept hoping . . .

Michel was growing frustrated by his attempts to contact the Josés. Francisco had completed his contributions, he declared, and everything now depended on the damp course and the *chape de béton*, the concrete foundation for the tiles.

Eventually, after a dozen or more calls, the gravel-voiced member of the quartet arrived, *petit* José.

'What are you worrying about?'

'We are into the third month, José.'

'We're waiting for a quote from the damp-course company. They're holding us up. This job'll be finished by the end of the week.'

'But we've approved the quote.'

'That chap's no longer available.' And so it continued until calm-of-demeanour Michel was a hair's breadth from losing his temper and threatened to cancel all outstanding monies. Then action recommenced, and at double speed. After four days of whacking, belting, radio blaring, concrete-pouring, tile-laying, the contract was finally completed, the site cleaned up and all was impeccable.

While they were charging their lorry, I asked why Francisco had quit and the small fellow went half berserk. 'Is that what he told you? Well, good riddance to him with his arty-farty ideas. We're the ones who do the work! He only drinks. Good riddance! We don't want him! He's not like us!'

I slipped niftily into the house where Michel was pouring us a glass of wine. 'The work is excellent,' he said to me.

I thought it resembled a jumbo-sized cardboard box, but this, everyone assured me, was what *gros œuvre* was about. The next stage involved an architect, drawings, planning permission. Once that had been achieved, the Portuguese would return to demolish the back wall in my den. Then construction of two upper-floor bedrooms would get underway. But those plans were definitely for another year. Michel and I were down to our last euros. High summer, July, was upon us, the empty cottage (Quashia's) needed roof repairs and guests would soon be descending. All demanded feeding, entertaining.

Extended days, languid and light, and the first of the oil bodies returned! The mutuality. A letter informed us that they were ready to reinstate us, on the understanding that we a) reimburse them the fees accrued during our exile, and b) sent proof of our new 'green status'. Michel attended to the matter directly.

I, meanwhile, was keeping a vigilant eye on the olives, plumping out, vibrantly green, flecked with white, but it was early in the ripening season. Still, the fly had not yet penetrated. Into this mood of optimism came an email from the same individual at the American chemicals company head office whose French technicians I had met with earlier in the year.

We were just wondering how you are getting on? Is your harvest looking promising? Are you using our product? Please remember, we want to do whatever we can to assist you in a bumper crop.

I was perplexed by this ongoing interest in our groves and I sent a return note directly. *The product is dangerous to insects, particularly honeybees.*

My one-liner was intended to put an end to this exchange.

That evening, true to ritual before the sun disappeared behind the hills along this Mediterranean coastline, turning the sky into a palette of scribbled crimsons, we were cradling chilled glasses of rosé – '*Santé*. Here's to the return of the first oil body' – when, suddenly, tranquillity was destroyed by the outbreak of barking.

'Not that Rottweiler, please!'

'I haven't seen that drunken neighbour in a while, have you?'

'No, but someone's been busy on their land. They have razed every single tree.'

Our dogs were now gathering at the foot of one of our centenarian *oliviers* down in the lower groves and no amount of whistling would persuade them to come to heel.

'I'll go.' I placed my glass on the table and descended by the stone staircase to find out what the disturbance was about. Something within the boughs, someone's cat perhaps, was agitating them. Or a red squirrel? I hushed the dogs and squinted into the silvery foliage where necklaces of early season fruits smiled down upon me. There a curious sight. I beheld, hanging from one of the higher, not-so-accessible branches, an old-fashioned string shopping bag. Or was it? About eighteen inches long, it was black, dense and . . . alive!

I shouted to Michel. 'Come quick!'

A feral swarm had set up home in our olive tree.

'Bees, oh, let's keep them!'

The problem was that we had no hive. I ran inside to telephone François for advice, but was shocked to learn from Marie-Gabrielle that he had undergone a quadruple bypass. It had been a life and death scenario. Although he was now out of hospital, he was too weak to come to the phone.

'Telephone *les pompiers*' was her advice.

I did as she bid.

The local fire brigade is much in demand in this Provençal climate of hot, arid summers and often uncontrollable forest fires. Still, within the hour two young men, fit as Greek gods, smiling broadly, chic in navy uniforms, appeared on the scene. They donned professional beekeepers' costumes – masked headdress, gloves and white bodysuits – while assuring us that the evening was the best time to capture wild bees because they were likely to be heavy with honey. This rendered them lethargic, more docile.

The darker haired of the pair, Denis, ascended our wooden harvesting ladder and began snipping with secateurs at the tree. I was horrified, watching on as he trimmed sprigs heavy with burgeoning fruits. Pierre, his companion, explained that this was essential in order that the capture went smoothly. Pierre, with open box in hand, now climbed a step or two and stood beneath his fellow *pompier*.

They spoke fast in twangy Provençal accents.

'Tu est prêt?' (Ready?)

'On y va.' (Let's go.)

Denis shook the branch and the swarm fell heavily in seconds. It reminded me of a stuffed nylon stocking. Wallop, into the box. Quick as a flash, Pierre secured the lid. A few little female bees had escaped and were now flying in circles, disorientated, but the operation had been handled skilfully.

'Félicitations!' I cried. 'How many bees?' I whispered.

'Ooh, 'bout ten thousand I'd say.'

The men hurried to their van where, in the rear, was a makeshift hive into which the bees were immediately settled.

Both the firemen refused any refreshment or remuneration. 'All in a day's work,' they grinned.

Denis confided shyly that his father was an apiarist. Alas, his hives had been killed off, and, if it was fine with us, he would offer this swarm to his dad who was suffering from the loss.

But of course. I was disappointed, sad that we were not keeping them but I certainly did not begrudge this man's father a replacement colony.

They patted the now docile dogs and wheeled off down the hill.

'Bonsoir! Merci beaucoup,' we waved. For one fleeting second I was glad that Quashia was not here to see them go.

In southern France, *les pompiers* are volunteers. They fight fires, they act as an ambulance service, they unburden the chimneys of nests, they rescue cats from treetops and, here again, they had proved themselves our local heroes. 'Such a pity not to have kept the bees for ourselves, not to have found a possibility of housing the swarm.'

We were returning to the terrace, to begin cooking. In our absence, while we had been out in the grounds, a message on the machine. Quashia had telephoned. 'Incredible. He must have known what we were up to!' I called to Michel excitedly.

He had left an Algerian number, but when I tried to phone it a woman's voice in Arabic, followed by the same in French, informed me that the number was erroneous. I tried every configuration of the ten digits he had given but none connected me. Telephoning Algeria had never been the easiest of exercises. I knew it from my own travels there. Frequently, it was little more than a game of chance. The fact that I understood this did not alleviate my disappointment. Might he be returning? Would we ever see our loyal friend again? If something were to happen to him, who would inform us?

It was eighty thirty the following Saturday morning when the full pack, a trio of Josés along with Francisco, I was surprised to remark, all laughing, joking, full of high spirits, came to collect their final payment, most of which was in cash, and sign off the contract. Michel offered coffee, which they all accepted save the little fellow who insisted on *vin rouge*. He said the wine helped him to grow! 'Never give up hope, *santé*!'

Michel felt that the work achieved by these crazy, disorganised southerners was not only satisfactory, but extremely fine. So much so that he requested a quote for the repair of the cottage roof. Unfortunately, they were not available. They had moved on to other projects.

'We could do it for you in October or November.'

We were keen for it to be restored as soon as possible. If Quashia were to return and find that it was still not waterproof, he might turn tail and fly straight back to Africa. We shook hands, compliments once more from Michel on the quality of the work, with plans to reconvene the following spring for the major construction work on the bedroom extension while, in the meantime, we had a dry garage and a spanking new private patio beyond our bedroom. But for the cottage, we were obliged to look elsewhere.

Alexandre, the handsome hunter, who worked at the garden centre, offered to take a look at the leaking roof when I asked him if he could recommend someone for the reparations. He arrived with his stepfather Jacky, who always referred to me as 'Madame', to estimate the damage. I knew both these men, had taken a hunting trip with them, and I had visited Jacky's home in the mountains where he lived with Alexandre's mother. They shook their heads as workers do here when they are about to quote an extortionate figure, but these two would never cheat me because we were, if not friends, then certainly more than passing acquaintances.

'The entire roof will need to be removed and replaced' was their diagnosis.

'But it is only a few cracked tiles!' was my retort.

'No, the damage is deceptive. The fabric of the ceiling has been destroyed by the incessant infiltration of water.'

If I had not known them, had not trusted them, I would have taken this for another example of local *escroquerie*, swindling, but the quote they returned struck us both as extremely reasonable. They requested access to our account at the builders' yard, to 'save troubling me every time they needed materials'. This provided, they set to, ripping, stripping, sawing planks, constructing a small scaffolding. Fortunately, the cottage was sufficient distance from the main house for us and our guests not to be disturbed by yet another round of construction hostilities. Jacky and stepson ripped off the old roof and laid the new one. They decided against putting down any lining; it reduced the expenditure and, in their humble opinions, the old Provençal method was perfectly adequate for such simple living quarters.

'Oh, but it must be watertight and lagged throughout,' I insisted.

'Oui, Madame, oui.'

We left them to it. The work was serious, elbow-grease fast and industrious; none of the cork-popping antics of our Latinos. Alexandre and Jacky completed the cottage within ten days: new roof tiles, freshly insulated, painted white, clean and neat throughout. We handed over the cash without question, grateful that this labour had been negotiated within record time, inexpensively and without hassle. Empty until the return of dear Quashia. His attempt to contact us had given me hope.

Two brief but intense mistrals swept through our broiling midsummer, but they failed to shift our mettlesome fruits. The olives were holding and there were still no tell-tale signs of female-fly penetration. Whatever we were doing right – I hoped it was the whole, the encompassing philosophy, but I could not be sure – it appeared to be working. The flies were not blighting and the fruits were winning. There was little else to do, aside from holding the fort without our man, keeping the grounds tidy and the monumental watering sessions; pool cleaning was Michel's task!

It was holiday time, time to take pleasure in summer, and the commencement of the non-stop traffic of guests. While they sunbathed or played around the pool, M and I remained in the sailed shade of the Bedouin Bar. I was engrossed in my Olive Heritage Trail, but I had discovered new reading, a young science: land management, stewardship of the earth, the creation of flower-rich habitats, environmental planning. I was in my element, trawling through science journals, blogs written by amateur enthusiasts, essays, programmes from agricultural institutes, learning of the natural balances that were set up among the various insect communities, the nesting sites of bumblebees, the plants they required – vetch, clovers, trefoils – understanding the vital necessity of hedgerows . . .

Armed with a large magnifying glass, I began happily rooting through envelopes of land, bent on the identification of one insect or another. Guests' offspring trailed behind me, the Pied Piper's Aunt. An unruly mob they were, screeching at a yellow and black 'flapping thing' or a 'green-armoured bug', but of all my companions of discovery, none took to the role quite like Marley. Vanessa and family were unable to visit, the same story with Clarisse, but Michel had persuaded his daughter to send her eldest son. This was his first experience of the Olive Farm and it had got off to a rocky start. The cluster of people round the pool, diving, jumping, squealing, sent him into overactivity, irrepressible weeping and wailing. One evening, his second with us, I think, I encouraged him along on a mammoth watering session, tears and torment blotching his sun-torched skin, introducing him to plants, pointing out the ballooning watermelons, where, to my horror, he jumped up and down and deliberately squidged one, but the scratchy call of the cicadas silenced him and he listened mesmerised as I explained the life cycle of these rarely visible little critters who did not stridulate but were always 'singing'. The owls' hootings, too, entranced him.

'You are right,' I said to Michel that night as we were drifting to sleep. 'He has a fascination with creatures.'

We assembled for lunches, dinners; an endless carousel of eating, laying tables, dishwasher-packing. Into raucous preparations while I was elsewhere on bumblebee detection with Marley and a gentle cousin of his from Cologne, a ringing telephone. Marie-Gabrielle.

'How is François?' I feared the worst.

'Poorly, but fighting, but I have delightful news to impart,' she exclaimed. 'We have a friend, Robert, who would be interested in placing a few hives with you. Are you still looking?'

'Indeed we are!' Bees, yaay!

Blue globes, Marley's eyes, bored into me.

The proposed beekeeper lived in St Paul de Vence, a destination not a million miles from our homestead, and more convenient than the mountain residence of our dear friends.

'We will be lunching with him at the end of the week. What say we drop in afterwards and introduce you?'

As soon as I learned the news, I was recceing the garden, seeking suitable spots, preparing for our future flying guests. I drove back to the garden centre to buy yet more plants, because, as François had accurately pointed out some time back, lavenders only feed the bees for one season and they were at their tail end now. Diversify!

Late that Friday afternoon, I was alone at the house – Michel had driven his sister, her two sons and Marley to the beach – when I heard the diesel engine approach and park. The apiarists were later than promised, unusual for them. I ran downstairs to greet them, shocked when they stepped from the car, Marie-Gabrielle at the wheel, François, bent and curled as a drifting leaf with a visible hand tremor and consistent head shake. His confidence, optimism, had seeped out of him, evident from his uncertain steps. Marie-Gabrielle led her love by the arm, assuredly, generously, to the table beneath the magnolia tree while I hurried inside for coffee, water and the uncorking of a bottle or two of rosé.

Robert and I left them to early evening refreshments, an amicable get-together in anticipation of a magnificent sunset, while we set off, strangers comfortable in one another's company, strolling through the dried tussocky grasses, the dogs running freely, looping in and around our steps. He rejected the citrus spot where François had placed his hives, so I led him up behind the recently established *Lavandula* bed, the spot Quashia had suggested months back. These terraces, I displayed proudly with their handsomely reconstructed walls, were where Michel intended to reinstate vineyards.

'Let's set the hives here, then.'

And so it was decided.

Robert oozed confidence. He must have been very handsome when he was younger, with the calm, weathered features of a yachting man, and bragged with a modest grin that he was the proprietor of several vineyards within the Bellet region in the hills behind Nice. His wine carried its badge of honour, its certificate of agricultural excellence, its Appellation Bellet Contrôlée. Seated back at the table, the conversation then turned, inevitably, to bees. French honey output was suffering, they said, particularly in the intensively farmed sunflower regions.

'But not,' butted in François, marginally revived, 'in the mountainous zones where chestnuts are farmed. If we had new hives and kept them at home—'

Marie-Gabrielle rested a caring hand over his, posed, shaking on the table. 'Hush, my love,' she said. 'We're fine as we are.'

I was disappointed that Michel had not yet returned and I tried to persuade our guests to stay for dinner but Marie-Gabrielle was concerned about the long drive home. Robert agreed to return with the first two hives after the weekend. Although no business arrangement had been discussed between us, I did not doubt that a few jars of honey would be offered and I hoped that he might also encourage Michel to get to work on the revitalisation of our long-defunct vineyard.

Monday dawned, steamy and humid. It had been raining all

night. One of those dramatic midsummer storms that light up the sky and beat frenziedly against the earth, drenching the crumbling sods. The church bells in the village were chiming eight as Robert chugged up in a forest-green four-wheeler. In the rear were two hives, one half-sized, the other the full quota. Sixty thousand honeybees making their debut at the Olive Farm. The dogs were running in excited circles around the Landrover while, in the house, Marley slept on. Out stepped Robert in bright yellow boots, ruffled hair and tanned, creased face. He looked as though he had just got out of bed.

'Bit late, sorry. I took the back lanes. Didn't want to transport bees on the motorway.'

'You're right on time. Welcome.' We embraced like old friends. 'Michel has taken his sister and nephews to the market in Nice and then the airport. He sends apologies, is sorry to miss you yet again. Yesterday, we levelled the earth and raked off the small stones on the chosen section of terrace. It'll make it easier to place the boxes.'

Robert was unlocking the boot. A quartet of stray bees were circling the car's interior and buzzing round the hives. To my amazement our new beekeeper squashed and killed three with his fingers. One was still crawling up the interior leather.

'Please don't!'

He looked my way, laughed. 'There's another sixty thousand in there.'

'Even so.'

He shrugged, began to unload the lighter of a pair of flat silver boxes. 'If Michel is not here, I'll need you to lend me a hand.'

Robert pulled out a carrier, containing wine. 'Try these. Two bottles produced from our vineyards up behind Nice. If you like our juice, there's plenty more.'

'Thank you very much.'

Between us, after the *chapeau*, the outer lid of the hive, had been lifted off, we carried the box of honeybees across the tiled terraces alongside the pool to the Second Plot, situating them above the extended lavender bed. Two metal stands were placed on the ground. Robert pressed them firmly into the earth until they sat at an almost indiscernible angle. One for each hive. Their slope allowed for the rain to run off. The dogs followed, tails erect and wagging. What could they sense, I wondered. Life, movement within, the ambrosial scent?

We returned for the forty-thousander and positioned it a healthy distance from its neighbour. Before he could release the bees, Robert needed to 'suit up' and he disappeared off, yellow boots softly sinking into the long golden stalks of dried-out grasses. How delighted Quashia would be to find these hives here. I was taking photographs, hearing his imaginary words: 'Has he brought honey? How much does he want for it? Will he give us some?'

Robert, masked and hooded, returned. Still snapping pictures, I maintained a wary distance while the flight paths were opened. I knew the bees, as they left their pads, would be agitated, prone to sting. Our new beekeeper ceremoniously, cautiously, slid open the hatches then rose from his haunches and crept stealthily backwards. The girls began to exit immediately, circling above their hutches, reorientating themselves, calculating their new location. Robert brought to mind a cartoon figure with his head orbited by dancing insects. He moved slowly, stepping twenty metres back, leaving the hives clear. Those that were already with him stayed buzzing round his head. The dogs started leaping and jumping. I called Homer to heel, fearing he might have been stung because he was snapping his teeth, trying to capture a bee in his mouth, but all was fine. I withdrew several more metres, backing up towards the tiled terraces away from the grassed areas, signalling the dogs to accompany me. Our new beekeeper had retreated in the opposite direction and was leaning tranquilly against one of the repaired walls up near the ruin, the ancient goatherd's hut. His girls began their inaugural flights about the land, zipping beneath the oak and olive trees, cautious of venturing too far from base.

Lola was charging endlessly round an outdoor dining table as though she had gone berserk. Suddenly, I saw Marley in pyjamas, standing at the far end of the table watching us, stepping to one side as the dog shot by him. I signalled him to keep his distance.

'*Calme-toi*,' I whispered to Lola and then a bee landed on my throat. I was in a V-neck T-shirt. The honey girl was treading across my bare flesh. I had purposely chosen pale blue, knowing that honeybees are disturbed by black. I stood stock still, letting the camera hang loosely at my side, knowing that if I remained calm, motionless, the bee would eventually fly off, but I miscalculated. Something spooked her. My breathing? The dogs' loony behaviour? Marley nudging closer? I remembered then that I had sprayed on perfume after my morning swim and shower.

I was not afraid. She stung me just the same. I felt the pain penetrate deep into my neck, but I stood my ground. If I flapped and knocked her away, her barbed stinger would remain buried in my flesh, and she would be defenceless and die. The same result if she panicked and tore herself away. Leaving the stinger results in the bee's demise. Better to let her extract it in her own time, which she did.

Eventually, we returned to the car and Robert disrobed himself of the apiarist's suit. Marley was at our sides.

'Why is he wearing those clothes?' the boy asked.

'We have bees, Marley.' And then to Robert, 'Coffee?'

'Sure.'

'I'll be upstairs when you're ready.'

'Are you fine?'

I nodded. 'I got stung.'

'Steep some cotton wool in vinegar and place it over the wound.'

'Can I see the bees?' begged my grandson.

'I'll walk you over there a little later, sweetheart, and we'll watch them flying in and out of their houses, but we cannot go too close or we'll frighten them.' At my behest, he hurried off to dress, thrilled by the prospect of a bee adventure.

Upstairs, I sat with Robert at the oak dining table, mugs in hand. It began to rain again and was falling fast and hard outside. One of our house guests must have filched the cotton wool from the bathroom and not replaced it. I was now wearing a crêpe bandage like a kerchief round my neck, soaked in cider vinegar, and probably smelt disgusting. We watched the heavy drops through the window for a moment, sliding against the panes leaving filmy snails' trails. Robert turned to me and smiled.

'Our weather patterns are becoming almost tropical, don't you think?'

I nodded.

'Sure you're OK?'

I gestured my reassurance though my neck was stiffening up.

He was talking in thickly accented English, apologising for the fact that he was hoping to use me as his 'practising partner'. I felt completely at ease in his company even though we were strangers. We had François in common as well as our mutual love for bees. He was seventy-six, which took me by surprise. I would have guessed mid-sixties. Returning to French, he began to tell me about himself, about the great famine that had driven his grandparents to the city.

'A great famine like the Irish Potato Famine of the 1840s?'

More recent. The famine of Savoie. Early twentieth century. I knew nothing about it. His grandfather with parents had moved west and settled in Lyon. The grandfather had started a small grain business. It became so successful he was not obliged to serve in the Great War of 1914. Instead, it was his duty to supply the troops with rations. Robert, the youngest of five siblings by almost two decades, was born and educated in Lyon. His three sisters and a brother had brought him up. After university in Paris, he travelled south to work as a marine engineer. Here he met his wife, a local girl from St Paul de Vence, which was where they resided to this day.

It was the tradition that every member of the local families up behind Nice owned a plot of land where they grew vines, the Bellet label. His in-laws farmed their allotment but without great industry or skill. No one took interest in it. They judged labouring demeaning. So Robert, while still building his career as an engineer, took over the reins of the family vineyards, buying up tracts from cousins and others who preferred banking to the sullying of hands. Today, his *vignoble* remained modest – 1.8 hectares – producing between five and six thousand litres annually. His son had caught *'le virus'*, the bug, and intended to take over the business. Between them they had expanded it into an award-winning, full-time occupation.

I reminded Robert that the terrace where he had settled the hives, as well as several others encircling the ruin, had once produced wine.

'Michel is very keen to reinstate a vineyard there.' This was a tiny moment of dissemblance on my part. I hoped my new friend would offer advice, would guide Michel in choosing the appropriate *pieds*, vine stock.

'Five thousand plants will be required to plant up a hectare and that will possibly deliver three thousand bottles of wine.'

Far in excess of our needs!

'Do you spray your vines?'

He shook his head. Although the vineyards were not situated by his home, being further inland and higher up the mountain, he was conscious of the dangers to man and pollinators.

'We won't be spraying the olives so, given that our neighbours are not farmers, the bees should be safe.'

'Who knows?' he said as he kissed me *au revoir* on both cheeks. 'Today there is always risk. Who knows what any neighbour is up to? Few are mindful of their responsibilities towards the flora and fauna. Most of us abrogate responsibility.'

After Robert had driven off, when the rain had eased, I returned with Marley to the terraced grove where the bees were settling in. The hives that had been left with us had both been swarms. The first had separated from one of Robert's own colonies the previous year, while the second had been a wild swarm. Robert had discovered it in the springtime, housed in the crook of a tree trunk on his land. I remembered our discovered girls and wondered how the fireman's father was getting on with them.

'When a foraging bee discovers a rich store of pollen, she communicates her find to the hive by performing a waggle dance. Her companions are given a sniff of the hoard she has returned with and then she begins to vibrate, shaking her abdomen, transmitting complex instructions. The direction the bee faces while dancing indicates the direction of the food source in relation to the sun. The length and intensity of the movement communicates the plants' distance and quality of food.'

When both Michel and the beating, sultry heat had returned, we sat out like picnickers on the evening grass with Marley with tray, lemonade and beakers, observing the bees return to the hives after a day's feeding. Beyond were the setting sun and the evensong of the birds. Marley and I had spent part of the afternoon creating a watering hole for the bees and then observed them alighting to drink.

He was transfixed. It was a battle now to prevent him from pulling the lids off the hives to look inside. An impulse I have frequently fought with myself!

'When will the man be back to take the lids off?' he wanted to know.

'In three weeks.'

His face dropped. He was returning to Paris.

Before he left, we took him to Nice and bought him illustrated books and reading material. Everything we could lay our hands on to immerse him in the worlds of what he described as 'bees and beasties'.

The boars (our beasties) had stolen the greater part of the apple crop and had cracked branches into the bargain, but the figs were ripening fast. Many had fallen, blotching the approach to the house with their squashed, bleeding seeds, but plenty remained from which to make jams and liqueur. The prospect for our olive harvest was looking promising and I was finally allowing myself the thrill of excitement, expectation. There was a percentage that had now been blighted but not the majority. I could not explain why we had not been plagued because the word on the olive network was that they had been out in force during summer. It seemed that we were blessed. The organic way looked as though it would pay off. I tried again to put calls through to Quashia but, on every occasion, the same message. We badly needed his industrious hands and without him we would have to seek assistance elsewhere. Together Michel and I wrote out the digits I had scribbled down and we tried every permutation, but still no luck.

It was the end of September, glorious golden days. I was in the garden collecting what remained of the apples when I heard the phone ringing and scooted into the house.

'Allo?'

'Quashia!' he bawled into the receiver.

'Mr Q!'

Even before a greeting, 'What's happened to the olives?'

And I was able to announce with pride and in all honesty: 'We have a crop. They are fattening nicely. Will you be back to pick with us?'

His brain was ticking over. The idea of missing a harvest, our harvest, his oil, too, the trees *he* had pruned, I felt sure was more than he could bear.

'I have no teeth.'

'The roof on the cottage has been replaced. No more leaks.'

'You only had to fix the broken tiles. I could have done it myself. Have you sprayed the olives?'

I could not lie to him. 'We've been keeping a vigilant eye and they are free of infestation. And, Mr Q, we have bees. Sixty thousand honeybees await you.'

His wheels were turning. 'I'll chase the dentist and find out how soon the dentures can be built. I'll be in touch', and with that he was gone. Neither a firm commitment nor a refusal. I sincerely hoped that I had hooked him.

'Give him time,' counselled Michel.

In the meantime, we commenced our annual preparations – laying of nets, cleaning crates. I telephoned the *bio* mill we had chosen in Saint-Cézaire-sur-Siagne, a medieval village rich in history, perched three hundred metres above sea level and one of the oldest olive communities in this corner of France. They confirmed that they would be open.

No more news from Q. I drove to the settlement and asked a couple of old codgers in long shirts and caps.

'He'll be back the day before yesterday,' cackled one.

'If not the day before yesterday,' his companion chipped in, 'it'll be the week after next.'

'If you speak to him, tell him, please, that we are waiting for him.'

And on another occasion, from a fellow sitting on a big boulder in the lane, all gums and tortoiseshell glasses: 'No, he's never coming back. He's gone to Mecca.'

We had to wait it out and trust, put our faith in Allah, as his companions advised me.

Towards the end of the month, our able-bodied man responded to my messages.

'How are the olives?' he yelled down the receiver. 'And the bees?'

'All is fine here, Mr Quashia. Are you coming?'

'The dentist has run off with my money and I have no teeth.'

'Oh. Could you harvest without teeth? I could introduce you to someone here. Remember, I mentioned a dentist—'

There was a crackle and then silence at the other end of the line. I thought we'd been cut off, until I heard him hollering as though into a bucket. His tinny response was 'I'll be there next month. I'll let you know when.'

'Looking forward to it, Mr Q,' I returned, but he had already gone. I was shouting into the watchful silence of the Algerian desert.

A few days before our staunch assistant was due to arrive, it began to rain. During this season, this was not unwelcome news. Olive farmers are gladdened by it. It fattened up the drupes and, when it had stopped, a kind bout of late autumn sunshine to dry out the fruits and branches, to prepare the ground, was all that was required. However, when the rains did not stop, that was when the farmers began to worry.

And so was the pattern for this late October. It began to rain, it continued to rain. It did not cease. When fruits were healthy, they withstood climatic changes, except hail. Severe wet and windy storms and they would usually survive but hail was the enemy at this time of year. Hail, on our farm, was a very rare sight and I was praying that this year would not prove the exception. A message came through from a friend to say that there were storms inland and they were travelling south. I watched from the windows blotched by raindrops. Miraculously, those storms did not come our way. They made a right turn and veered off, whining and screeching, into the higher reaches of the Var, but the rain did not follow the hail. It stayed with us. Relentlessly falling, soddening all.

And then it stopped.

I made a tour of the groves. The majority of the olives were hanging on to the branches, mottled and fat.

'We're getting there!' I yelled into the damp air. Puzzled expressions from the dogs.

Then I received the call from Marseille – 'Quashia! I'll be on the five o'clock train.' I drove down to Cannes to the station to collect him. He had arrived with two of his fellow countrymen, shuffling, scruffy Berbers from the same village. He had gained weight, his paunch fattened by the cooking hands of his wife and a clutch of loving daughters and daughters-in-law. His navy jacket lay untidily open. Strutting like a pigeon, his black lambswool hat crowned his balding head. When he smiled I saw that he had a full mouth of teeth.

He shook his head. 'They're giving me gip. I wish I'd kept the rotten ones. Might we give my pals here a lift, save them a taxi? They travelled over on the same flight.' He spoke with confidence, the leader of the group.

'Of course,' I smiled, heartily embracing Q and handshaking the others. The luggage for these three men, all returning after months in Algeria, consisted of four stuffed-to-bursting plastic carriers, one carton-shaped tartan bag and two cardboard boxes tied with string. One of the carriers, torn, contained a generous swag of fresh dates. These were a gift for us, for Michel, who adores them. Usually, they are picked from Q's backyard moments before he sets off, but on this occasion they had been purchased at Constantine airport and bore a little tag with the price still on them.

'I didn't have time,' he apologised, 'and I didn't want to arrive empty-handed.' His habitual generosity had not abandoned him.

In the car, he sat alongside me with his compatriots in the rear; shyer men, less at ease in my company, a Western woman. He was ebullient, full of high spirits at being back. He was calling over his shoulder to the others – they might have been passengers in another vehicle, he shouted so loudly – that he had a great *patron*, boss. 'Who else would come to the station to meet us?' he laughed.

I was pleased to see him looking rested and in such a breezy frame of mind. There was no trace of his previous upsets, no mention of the long months since we had seen one another. Perhaps he had missed us as we had him?

As we climbed, winding up the hills, the men remarked on the dark red earth, the rich colours of the foliage. All fed by rain. I admitted that we had suffered several days of storms, heavy downpours, but it had cleared now and the sun was shining, fattening the fruits, encouraging the oil within them.

We deposited the men at their lodgings lying in the valley beyond us, and then I returned Quashia to his gate. Our cottage.

'There's fresh coffee, a new percolator, a *baguette*, some cheese and veggies waiting for you,' I told him. All these save the *baguette* and veggies I had prepared the previous day when I had looked in on the place, opened up the windows to air it, confirming that the new roof was intact. I would never buy meat for him. Following the edicts of his faith, he ate halal, purchased down in the Arab quarter in Cannes not far from the station, but I left that to him. Once I had deposited in his fridge four delicately pink lamb chops from our butcher. He had not been pleased.

'There's no meat, of course,' I grinned as he hauled his luggage from the car.

'I'll eat with the others in the valley.'

'Want me to take you back there?'

He shook his head. 'No, I'll stroll down when I'm ready. I'll see you in the morning,' he winked. 'We'll get the nets out and give the grass a quick cut to prepare the ground for them. When does Michel get back?'

'Friday. The nets are out, the forecast's good.'

The days that immediately followed were ambered with a warm autumn presence that breathed softly through the trees and rested beams of light across the trunks of the great pines up in the forest near the summit of our hill. Quashia and I worked out of doors, each of us in his own world, playing our part towards the ancient rite that lay ahead. When Friday came, I collected Michel from the airport and we settled to an evening together in front of the big open fire, recounting to one another the news of our separate weeks. When either one of us, or both, had been away on a working trip, Friday evenings took on a special meaning. Its mood of settling back into the rhythm of one another, of being at home, shedding the stresses of the city and opening one's heart once more to the home, the farm with its silences and its familiar perfumes and sounds.

'Marley wants bees and a gecko,' he laughed. 'Driving poor Vanessa insane about it.'

Oak and olive logs crackling on the fire, the deep bass lick of the flames consuming the wood, owls hooting further up the hillside. We stood out on the terrace after dinner gazing at the stars, listening to the night. The dogs were sleeping. Lola, snoring softly. The sky was partially covered and Michel observed that we would probably be seeing a little more rain. 'Over there,' he said. 'See those navy clouds banking up, gathering strength.'

We did not need rain now. Sunshine was what was required

for the harvest. I had booked the first mill appointment for the following Wednesday. A few days of solid labour lay ahead for the three of us. We needed all the good weather we could get.

Saturday stayed fine. While Quashia took the morning off to go to his Arab market and stock up on provisions for the week ahead, I took yet another stroll, this time in the company of Michel, up and along every one of the olive terraces, both the new groves and the old. We logged the condition of the fruits and, in both our opinions, they were robust, ready. Some showed signs of fly infiltration, a few more than last time I had looked, but more than 50 per cent were turning nicely. They were ripening a little early, but that had been the norm for the past three years so we were not alarmed by their rich, blackening skins.

This would not be our heaviest load, but it would be an organic harvest.

'Remember,' said Michel, 'it's not bumper crops we are after, but quality food.'

Sunday, it began. Sombrous weather, louring clouds. In the quiet of late afternoon as darkness was falling on these shorter, smoky days when fires are burning in hearths all along the hills inland of the coast. Mid-November. The sky was as black as spilled ink. Five weeks prior to the solstice, before the sun was spinning to its winter standstill. I was boiling the kettle for tea. Michel was in my den tapping at his computer. I lifted my head, paused, heard a clap, a distant, sonorous warning. Moments later, an illumination lit up the sky beyond the back of the house. Thunder somewhere far removed in the mountains. I continued with the tea-making. Cups on saucers and on to the tray. Steaming water splashing into preheated pot. I carried the tray through to the *grand salon* and placed it on the low wooden table by the fire and picked up the book I had been reading.

'Tea's ready,' I called through to Michel who I noticed then for the first time had abandoned my desk, had unlocked one of the sets of French doors and was on the upper terrace that looks over the sea. Like a retriever picking up the scent on a shiver of wind, Michel was alerted.

'Rain's coming.' He stepped back inside and turned the lock in the door. 'It's promising quite a storm.'

I sighed. 'Just what we don't need. I hope it gets it over with tonight and the sun dries it off again tomorrow.'

But the rain had other plans and our olives were not a consideration. It reached our skies shortly before dinner. Commencing slowly, big drops that thwacked against the windowpanes, as though they had heard my dismissal of their arrival, my shoving the storm onwards too quickly. It was proving its force. 'There is no sending me away because it does not suit you', it seemed to be telling me. And then it swirled and danced. Rising up with a wind, skittering whorls of fallen leaves into the still, deep pool. And this was just the beginning. Thunder came, growling loud and angrily. A voice without reason.

'Oh, Lord,' I murmured as I served potatoes. 'Where are the dogs?' They were not in the house.

'They were in the *hangar*, sleeping by the wood stock last time I was up there fetching logs.'

Since they had lost their stables to the Portuguese conversion, the two dogs had not entirely settled, sleeping here and there, rarely in the same place twice though I had made concerted efforts to rehouse them. 'I'd better go and look for them.'

'I'll go,' said Michel. He crossed through the sitting room and unlocked the front door and the weather-stranger flew in, gusting, causing everything to quake and rattle. A wet tempestuous wind such as we rarely encountered here.

I was shocked, let fall the serving spoon and hurried after Michel. Out on the terrace, I stared seawards. It was sullen, dark, roaring, while the old olive trees, silhouettes this side of the sea, seemed vast, troubled, grieving phantoms.

I hope this lets up so that we can get to work, I muttered.

All day Monday it rained, gnashing its teeth and beating hard against our flat roof. Quashia did not come up to work. There was nothing to be done. I ran through the floods to feed the dogs, to find them and comfort them, to bring them in with us and let them bask and burn, dry their damp, smelly coats. There was nothing for us to do but wait, to stare out beyond the windows and be patient. Nature was the mistress here, brandishing its force.

Eventually it ceased, the rain. The sky lit up and the afternoon opened like a flower. It was magnificent. Tuesday. We had lost time and the mill rendezvous had been booked for the next morning. It was an impossible challenge, but we would not be the only farmers who had lost out. I strolled in the late-year sun down towards the lower groves, stretching my limbs, taking stock of the trees, silvered and energised by the rain but lacking their fruits. I looked at the circles around their roots expecting to find puddles of oily olives but there were few. A fleck here and there, nothing more.

I called to Michel, who came out on the upper terrace.

'Is Quashia with you?' he shouted.

'No.' I was descending the slope of land.

There they were, drifting in swarms, like dead locusts, blotching the property's perimeter. Hundreds of kilos of olives. I raised my head and stared up into the still sopping branches, rich with leaves but devoid of fruit. I called to Michel who was on his way down. Together we walked the terraces from the bottom of the estate to the very top, zigzagging, up and about. The paths, the grasses, the stairways were deep with fallen, ripening fruits. Every tree told the same story.

'We'd better gather what we can now and hope there are sufficient undamaged to make at least one single-estate pressing.'

A pressing of our own, a 'single-estate', required, these days, a minimum of one hundred and eighty kilos of good-quality fruit.

Michel was right. We had to try and salvage what we could. But where was Quashia?

'Perhaps he's sleeping and hasn't realised the rain has stopped. I'll get him.'

I knocked on the cottage door several times and received only silence as a response. It was puzzling. I telephoned Quashia's mobile from my own but it was switched off. After another try, a final loud rap, I pulled out my key and opened up. One foot inside, I felt the cold as my shoe sank into water. This was not simply a leak. The floor was a lake of fallen rain and floating upon it was the debris of a building and a life: papers, bits and pieces from the kitchen, furnishings, all were bobbing about on its grimy surface. The new roof had given way under the torrent.

But where was Mr Q?

I waded through the flooded living room towards the bedroom. The door was closed. I knocked. I was praying he was not in there, that a displaced chunk of plaster or beam had not fallen on him while he was sleeping.

'Larbi?'

Silence. I felt hot and cold at once, an appalling sense of dread. Slowly, I pushed open the door. The bed was a mess of soaked linen. The room, too. The wardrobe door was open, its interior empty, a few items of his clothing were sinking in the water on the ground. I looked for a body. There was none.

I rang Michel. 'I hope he hasn't just left us.'

'Don't be silly.'

I went looking for him at the settlement where his Algerian chums were and found him there, limping. He told me that he had packed up his bag with his personal belongings the afternoon before and begged a bed.

'What happened to your foot?'

'I broke it,' he moaned. (Quashia was a child when he got sick. Even a cold and he moaned that he was dying.)

I drove us both back to the cottage where Michel was chucking buckets of water out into the garden, attempting to dry out the mess. The roof was a catastrophe.

'I'm running Quashia to the hospital.'

'I'll call the Portuguese.'

Our desert man was surprisingly good-humoured as we sat in Emergency. It was a long wait.

'Those olives were blighted,' he said to me. 'Not all, but the majority. You hadn't noticed because they were changing colour. You lost them because you didn't spray.'

'There might have been sufficient healthy ones for a pressing.'

He shook his head, rubbing at his lower leg. 'Two or three kilos, what's the point? You'll only get bumper crops if you spray.'

His foot was not broken, of course. He would not have been walking had it been. He had slipped and sprained his ankle, trying to escape the interior flooding. No serious damage.

Back at the farm, we found Michel, sleeves rolled up, collecting what few fruits he could, but there was precious little to salvage. We had missed our mill appointment and the drupes were too wet, too bruised to keep for any length of time and Quashia was correct. When we examined them closely, many had been attacked by Dacus, but somehow it mattered less.

'When all these olives are sufficiently dry,' said Michel, 'they'll have to be burned.'

I nodded. It was a sad conclusion to our first year of organic farming.

My profuse apologies to Quashia were waved away with smiles. 'It'll get sorted. I'll stay where I am until the cottage is repaired. What matters is to get all this mess cleared up and get ready for next year.'

I loved him all over again for his consideration.

12

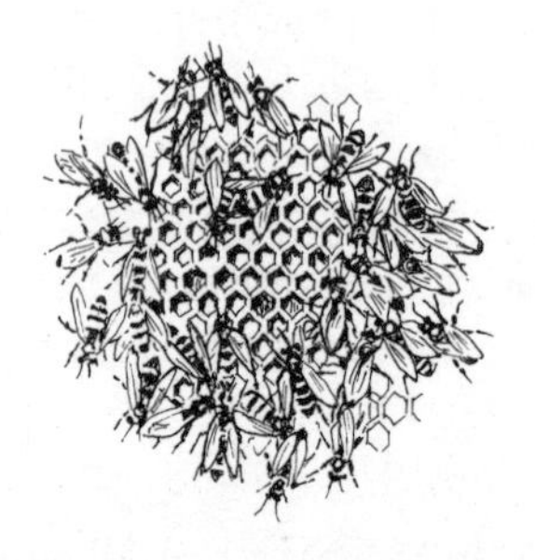

A quartet of wood pigeons were flapping their wings like sails in the descending light. Doves were cooing in the gloaming. The days were growing longer. Bumblebees, several species, were busy in the lavenders. A light mist sat on the heat. A few low cumulus clouds like a slick of creamy-white milk spilled into the valley beyond the farm. Others, into the distant *vallons* along the Estérel.

I was making my way carefully through the fields bursting with flowers, busy with insects, with Nature going about her business. In the words of Virginia Woolf, the garden was 'full of lust and bees' and it was spilling over with colour. The seasons, the months, were being reflected chromatically. In January, there had been a golden carpet of daffodils, narcissi. Fields, terraces buttered with lemon-yellow wild flowers. The fragile pink of the almonds in February. March had delivered linen-white blossoms on the fruit trees, including the junior cherries. Now it was May once more with its vibrant and varied tapestries, dominated by wild flowers, the rich red, pink and cream roses and the jasmine and orange blossom. The perfumes made my head reel. I wandered the terraces as though drugged – the grassy wilderness caressed my knees and I felt sure that any day now we would be receiving a letter from the council threatening us with a fine if we did not cut back the herbage. Already, there were risks, due to above-average April temperatures. The coast had suffered its first fires in the Var. No human casualties were reported but thousands of acres of forestland had been destroyed.

The stalks of grass had grown so tall now and so barley-green. On several of the terraces, they were brushing against my buttocks. 'Hey, lady, look at us!' I could no longer spot the dogs when we were out walking. It was the movement of grass, the tossing heads of the armies of purple thistles, the bristled and unexpectedly sharp stalks of the vivid viper's bugloss, *Echium vulgare*, that alerted me to their whereabouts. The grasses, wild flowers swayed softly, dislodged by the animals' bodies. *Echium vulgare* was a wasteland plant that had been found to contain significant levels of gamma linolenic acid and a rarer stearidonic acid. Like borage, which spreads profusely across our terraces, and evening primrose, it contained essential fatty acids. It was also a hearty contributor to the tree bumblebee's diet.

For many years I had been teaching myself the identities of the plants. This had been for no purpose beyond my own curiosity. To acquaint myself with where I was living and what other forms of life inhabited the hill alongside me. Now the interest had another purpose: I hoped to understand the delicate balance of the species' needs. All part of my new land management programme!

The tuberous comfrey had never reached flowering stage before this year. It had been felled. The terraces were scattered with wild Mediterranean orchids. In the tropics they are epiphytes – plants that grow on other plants. Not here. Here, several species of Orchidaceae grew directly from the earth, but we had never seen them before because they were levelled by the cutting machines before they could identify themselves. The red helleborine; broad-leaved helleborine that provided no nectar but offered a nectar-like wax for wasps. The pink butterfly orchid happily inhabited this *maquis* region and I had been completely unaware of it. All pollinated by wasps. And the bumblebee orchid, an inconspicuous treasure, whose existence I had been blithely ignorant of, was, like me, happiest when haunting olive groves.

In order to adhere to the regulations laid down by our organic status I was now obliged to resist, during my regular trips to the garden centre, purchasing half a dozen sacks of *terreau universel*, all-purpose earth. It was time to bring our own soil from this crusty limestone hill into play. With our pine forest high on the hill's apex, quality natural soil existed. Comfrey in the compost heap offered us an organic nitrogen, which was a perfect way to keep the soil balanced. Comfrey was a natural fertiliser of the soil.

A detail, one of many, each remarkable.

Every morning I rose at daybreak and hurried impatiently outside to gaze upon the high, rising grasslands. I did not wait to percolate coffee. I needed to be there, out among it all. The birdsong was intensifying. The farm was abounding in songbirds, butterflies (dozens of *Vanessa cardui*, which thrilled Marley to know that a butterfly bore his mother's name!) and bees. Among them was the European tree bumblebee, our own honeybees, of course (we now lodged over 200,000) and myriad species of other pollinators.

The reign of the fascist-like magpies was becoming a thing of the past.

I felt as though the ground was reawakening, unfolding, opening its eyes and grinning. When we first cut back this hillside, a monumental and very expensive task, hacking and sawing through years of jungled growth, I felt that the sods of earth, the dried-out soil buried so long beneath briared darkness, had lifted itself up, inhaled the rays of the sun, sighed, stretched its limbs, glad to experience its freedom once again. But today's was a different reawakening, or so it seemed to me. The earth was regenerating, feeding itself, calling upon its own brotherhoods to return, to take command.

I had only to step out of the way.

These were the final days of late spring, and the *Lavandula heterophylla*, sweet lavender, planted the previous year, was becoming a favourite hangout of so many pollinators, particularly our honeybees, which seemed to be fixated by the violet blossoms, swarming and circling the plants, working them for hours on end. On one plant alone, Marley and I counted twenty-five bees this morning, among them three tree bumblebees. They resembled the common bumble except these had white, furry backsides. I had only recently encountered these little darlings.

I had joined Apimondia, an international organisation representing the interests of apiculture, and learned that last year approximately 30 per cent of Europe's 13.6 million honeybee hives had died, been killed off, disappeared. Even the very largest businesses in beekeeping – apiarists who owned up to and over a thousand hives – were facing bankruptcy. This tragic news had one positive side effect. People from cities with smallholdings were taking it upon themselves to Save the Bee and were becoming part-time apiarists.

After his holiday with us last summer, Marley had returned to Paris impassioned by some of the creatures he had encountered here and he began to drive his parents crazy with his requests for 'beasties' of his own. I encouraged Vanessa to plant window boxes on their matchbox patio. Even this did not satisfy his eager desire to become acquainted with the natural world around him. It was not until, quite by chance, I learned from Robert that he had colleagues in Paris who were installing hives on rooftops: one at the Opera House, another close to the Eiffel Tower, a third on top of the Grand Palais. Chic bees! All these urban hives (more than three hundred in total) were flourishing while those maintained by the same apiarists in rural areas, agricultural zones, were ailing, if not dying.

'There has to be a message in it somewhere,' Robert said to me. 'People talk of mobile phones as the perpetrators of the problem, but there are hundreds of thousands of mobile phones in the cities and the masts for them.' Robert said he was convinced that it was the pesticides used on crops in the rural areas.

Might he agree to contact one of his beekeeping associates? Might we arrange for Marley to pay those hives a visit? It was then that I learned of an extraordinary story. Close to the Boulevard Saint-Michel, there is an apiary with but a handful of hives whose beekeepers are dedicating time to working with children, children with hearing difficulties. Their goal is, through the social network and colony structure of bees, to assist in awakening worlds for those youth who are otherwise partially closed off. A fact I had never even considered before: honeybees have no ears but they have an auditory system of sorts, a Johnston's organ at the base of their antennae, as well as a complex system of communication as evidenced by their famous dances. Michel and Vanessa took Marley along to visit the apiary, hooking the boy up with this remarkable world and its special forms of communication.

When I shopped, I shopped for an augmented family, the bees as well as the dogs, ourselves and our guests. I bought extra plants, always trying to think ahead, to be sure that when one species had completed its flowering, there would be others in blossom. The financial saving on insecticides was greater than I had expected. We had never sat down before and calculated what our old way of farming had been costing us. Not fortunes, but sufficient to splash out on the creation of gardens, transforming forgotten pockets of dust and stone, envelopes of land into extra herb beds, climbing corners, flowers creeping through and along the metres of balustraded terraces. Here was the purple bed. Elsewhere, the white corner. I had learned somewhere that the Bombus species, the bumblebees, were attracted to white flowers, and I observed that our honey girls frequented the lavenders and others with tall, spiky, purple-toned heads. And how I indulged them! I sought out flowers that I had never heard of before, returning with the back of our van loaded with myriad varieties. Quashia would shake his head and ask: 'Where's this to go then? What's this for? Is it a vegetable, a herb, can we eat it?'

'We cannot, but the bees will enjoy its offerings.'

He was happy. I would watch him through the window from my den, shadowed by the loyal dogs, taking cuttings, planting them into tiny pots and leaving them on handkerchiefs of soil where he felt there was insufficient growth. He had stopped nagging me about weed clearance. He was counting the months before his cottage cupboard (now renovated by the Portuguese, but at such an expense that we had been obliged to stay all architectural plans on our extension for a while longer) would be bursting with honeypots.

Robert, who was assisting Michel with the preparations for his soon-to-be realised vineyard, told me that Quashia had taken to phoning him. 'Anything I can do for the bees? Are they comfortable, happy with us? If you have a spare pot of honey, bring it for me when you next drop by.'

Numerous species of bees in the garden. I stalk them, in a kindly way. I identified two Blue Carpenters on the passion flower though I found no traces of their nests, even though we had logs and dead wood aplenty. And one morning, up near the woodshed after feeding the dogs, I came across a mammoth flying insect. Similar to a hornet, but this I had never seen before. Close to two inches in length it was, with four clearly defined yellow squares on its back. It was hovering above the grass and then settled on the ground and darted forth into the tiniest of cracks, disappearing underground. It took me the best part of a day to discover its identity – a French Scoliid wasp. He was a solitary who lived off the larvae of scarab beetles. Across the planet, there were three hundred varying species of this fellow.

Oh yes, I could feel the shift taking place within me. Giving up control.

I am simply a worker here, just like the foragers and pollinators, just like Quashia. I had a part to play but I was not the orchestra leader. Neither was Michel. Nature guards that role for herself. She does not enforce her superiority but she maintains it and in the understanding of that, in the yielding to the act of creation, I felt more content than I ever had done before. It seemed that for the very first time in my life everything was in its natural place. I was no longer hell-bent on attaining, achieving, rearranging. This order might be chaotic but it had not been imposed and my job was to respect it, to be kind to the earth, to assist with its perpetuity, not its destruction.

After the loss of last winter's olive crop, the three of us had worked hard. We had cleared the land, burned the damaged beauties, raked and tidied. Climbed within forests of branches hunting out the rogue fruits, those that had clung on, and we picked these too and relegated them to the bonfires.

I had no means of forecasting whether or not this 'new way of farming', this natural approach, as old as agriculture itself, was going to work. What I had learned was that I had little say in the matter. Nature did not need me, any more than the Olive Farm did. When I went away for a period of time or fell asleep, it did not stop living, breathing, working, creating. Nature was always busy, always weaving. I had a choice: hinder the process or leave it be. I also had the choice to be there with it, to create at its pace and not mine.

In the years since we have taken possession of this olive farm, I had seen a shift in attitudes towards farming, towards living. The world was spinning in confusion and little seemed to me to be real. Love is real, I think, and Nature is real. And we cannot own either; we can engage with them, play within their boundaries, draw immense energy from them but not much else.

There were steps to be achieved along this journey. No more urgent issue, it seemed to me, than to steer a new direction away from the pharmaceutical multi-nationals, the agrochemical giants who had the world within their grips. I received one more missive from the chemical company in the States. They sent me through pages of notes, experiments that were intended to prove that their product was not harmful to insects other than Dacus. I was not convinced. In any case, my return to them was: *Thank you, but we are not looking for 'bumper crops'. We are looking for a method that allows us to live with Nature not as its dominator.*

There was no easy solution. We had lost the crop, were walking on thin ice with the olive bodies and I saw no assured direction ahead. Less oil had been my expectation, my calculation. None at all, for a second or even a third time around – this I had not been anticipating. We had not eked out what we had, we were living and eating with our usual lust, gusto for food and the pleasures of the garden and yet there remained well over two hundred litres. That was more than sufficient for our needs for yet another two years. Either I had miscalculated the previous June or, like the parable of the loaves and fishes, the stockroom remained bountiful.

The earth would give us all that we required, I felt sure of it.

Marley had come to visit with Vanessa. We were out on the land, walking in the late afternoon light. Michel with his grandson took the lead – they were naming the trees, the fields and terraces – while Vanessa and I strolled slowly, bringing up the rear, chatting of inconsequentials.

'Look! See that small olive tree there?'

The blond-haired boy nodded, his uncertain eyes as large as marguerite daisyheads. He was a city child with urban images in his head.

'Many of these trees have names, Marley, and for those that are not so lucky, we can christen them. That one there is Bridget. She has a mournful soul. See how she bows her crown towards the elephantine fig towering above her?' The boy stared at his grandfather and then back at the growth. Slowly, with a certain amazement, he nodded.

'That tall one up there, way up high. He is Hans. He shoots skywards reaching towards the hill's pinnacle and he pays no heed to the shadows cast by the pines. They can't frighten him. He's a warrior.'

'I want to choose a tree,' shouted Marley, breaking away from his grandfather's grip, haring this way and that. 'Maman, which tree shall I choose?'

I caught sight of a song thrush in the garden cracking open the shell of a snail on one of the steps of the Italian staircase.

'Marley!' An urgent, soft call. The boy crept back towards me, watching us watching the bird.

'Snail,' he screamed.

The thrush took flight. Vanessa laughed and drew her son towards her. 'When you shout, honey, you frighten the birds away.'

This consideration had not occurred to him. He pondered for a moment, amazed at the prospect of such responsibility. Then he shrugged his body clear of his mother's embrace and went in search of the snail, bending to it, gathering its broken shards into the palms of both hands.

'Snail!' he giggled in an ecstasy of delight. 'Can I have him?'

'That depends on what you want to do with him, Marley.'

He stared at his mother and then at the morsels of dead life. 'Try to fix him.'

We found the tiniest of voles, Marley and I, keeping itself

cool on the top step of the pool where the gentle wave motion lapped against its miniature underside. When I went near, it dived like an Olympic medallist into the water, submerging itself, kicking small back feet, moving at a rate that astounded me. And another time, across the grass a russet hare with her leveret. Mother, stock still, waiting, and then they bounded off, slipping artfully beneath the fence, disappearing into undergrowth that belonged to our neighbour, the hunter. The boy at my side kept shush while the dogs took no notice. They did not even seem to have registered the quivering presences. Bassett, our little hunting hound, I recollected with a stab of grief, would have been a bullet on their heels, but I took great pleasure, comfort, too, in the company of this boy, in sharing his wonder at the living beings around him. I would have found great joy in teaching a child of my own and part of me yearned for him to be mine.

Our mightiest of pines was a great vertical being with branch spread only at its crown. Possibly the tallest tree on the estate, it bore the scars of its history. Several of its branches hung, ripped and torn, creaking gothically in the wind; callused witnesses of past storms. Limbs that would never heal. I had christened him 'The Aboriginal' because he bore a beaten brown trunk, a coat of weathered, crevassed skin, and he watched out to sea without a word and within his upper echelons tawny owls made their nests. I watched their comings and goings from our bedroom. The Aboriginal was a silent, damaged soul who knew the lie of the land, kept his counsel and was gentle and generous towards his nocturnal lodgers.

A loneliness, an isolation, that I think I have carried all my life was lifting, or perhaps I was accepting of it, no longer perceiving it as negative. Beyond 'Carrot', the little girl I never gave birth to, whose beaming smile still paid occasional visits, though less frequently now because I was letting her go – her unborn spirit had moved on to another dimension – beyond her I had found a family here, speaking the tongues of the world, the tongues of Nature, too. Footsteps fording through the fields, through the seas of growth, I encountered insects everywhere, grasshoppers, tiny triangular chaps in bright green that resembled miniature tanks, a rare sighting of a praying mantis, pop-eyed and graceful. I scratched my scalp and found a beetle with a brown posterior rumbling about in my hair. As busy as the earth was, so, too, was the air. Flitting with life, motes, seeds, variants of midges swimming about in the warm sultry days. Life was returning to the farm. Not a controlled human vision of what this should look like, but the life of Nature in all its glory and harsh realities.

I stood within all of this, this harmony and chaos, listening to it breathe, its whisperings among itself, sharing secrets between bird and bee and stalk and flower head, and a tear rolled softly, silently, down my bronzed, summered cheek. I had been foolish, unwittingly so. Foolish to wish that Marley was mine. This farm was not mine. It was not ours or mine or anyone's. I was simply a passenger on this swelling geoponic sea, sailing through for a time, dislodging the plants as I passed on my way. I was curator of the Olive Farm, an honourable role, but, once I was gone, out of sight beyond the horizon, having disappeared between invisible doors in the sky, drawn inexorably from beyond when my time came, through those blue-tinted doors, this hillside would forget me. It would continue as though my passage had been nothing more than a heartbeat, nothing but an evening sprint of the dogs bounding across the terraces.

And then?

Spring would return once more to the Olive Farm and the flowers, the buttercups and dandelions, golden orbs of the fields, all would bolt skywards with glad hearts, laughing at the extravagance of their own gay, irresistible colours. And beyond, after the shedding of those shy, delicate, white-laced olive flowers, a trillion, million tiny green nubs would nudge forth like dozy slow-worms, heralding the arrival of the olive, which would evolve, fattening up its ovalness until Dacus reappeared, to pierce those delectable fruits and lay its eggs. Deep within the chocolate-brown silky silence of the pulp, the larvae would hatch, feeding, sucking, guzzling, diminishing the fruit's life, destroying in its propagation the oleaginous being that was feeding it.

And even I, were I still here, could do little. Together, they must find their balance. Dacus had been my enemy; it was the olive's enemy, too, but to *Psyttalia*, for example, it was food, the next meal. A natural battle for survival. A symbiosis that had little to do with me, a rectification beyond the poisoning of the earth.

We had lost the map, the chaos that gave life to all things, that named no names. Is the earth's mystery slipping away from us? Have we lost sight of its power, its mythology, the stories that it carries in its belly and heart? The ever-unfolding narrative of a landscape, of the details of its topography, an exploration of its botanical components. History, literature, myth, folklore, botany, geology, language. A pilgrimage across the earth.

My life had become a world of small things, a story of minute details, of the earth's growing. I found joy in minute observation. It was about looking, seeing and hearing the sounds of the land. Salamanders or newts at the pool's edge, a pure white egg, broken, now abandoned, birthplace of a woodpecker or owl? Buzzards circling overhead in the columned blue silence of summer. Painted Ladies poised on bay tree sprigs, delicately flapping wings of arabesque beauty. Bats silhouetted against a heliotropic light, emitting their high-pitched keening, their echo-sounding.

Returning; the creatures were returning. The changing moments, from break of day throughout my waking hours into evening, night. My mind was growing quiet. It was growing still, accepting of all that was around me. Such quietude, flowing like mercury through the troubled spaces in my head.

The tragedies, battles, heartbreaks of former years had receded from view, from memory, but I knew that, like storms far off on the ocean, they would return in unknown guises. This intermission was a gift, a blessing to be treasured. To take simple delight in the yelp of a daisy, to bask in a rose's silken caress, the dew's song, the dawn's blues, the trefoil's tiny giggles. Lullabies or reckless screechings that shattered the sky's tranquillity.

The sky was bluer, more azure-deep, than I ever recollected it. Craning my head, peering up into it, gulls flying high, luminous in the bright midday sunlight, wheeling, turning, they cast sweeping black shadows across the olives' canopies. Feet on the ground, I stood gazing out to sea or spun and peered inland into the petal-soft lilacness of the distant mountains and I felt immeasurably happy. Yes, immeasurably joyous. During my travels, I had sometimes felt a stab of fear, anxiety at the prospect of returning. Dare I return, dare I? Returning is what they say travellers should never do. But I had had nothing to fear. This traveller had not returned to the Olive Farm.

Not at all. She had just arrived.

Index